ez101 study keys

Physics

Second Edition

Patrick C. Gibbons, Ph.D.
Professor of Physics
Washington University
St. Louis, Missouri

All inquiries should be addressed to:
Barron's Educational Series, Inc.
250 Wireless Boulevard
Hauppauge, New York 11788
www.barronseduc.com

ISBN-13: 978-0-7641-3919-2
ISBN-10: 0-7641-3919-3

Library of Congress Catalog No.: 2007935674

PRINTED IN THE UNITED STATES OF AMERICA
9 8 7 6 5 4 3 2 1

CONTENTS

Prefixes :   Modify units of distance, time,
                            change, etc.

Mega	10^6
Kilo	10^3
Milli	10^{-3}
Microon	10^{-6}
Nano	10^{-9}

DESCRIBING

MOTION

This theme presents the concepts used to describe motion—distance and position, speed and velocity, and acceleration. Scalars, which have magnitudes and may have dimensions, are distinguished from vectors, which also have directions.

Scientific notation is frequently used throughout these keys. Very large and very small numbers are written as 3×10^8 or 1.6×10^{-19}. 10^4 means $10 \times 10 \times 10 \times 10$, and 10^{-4} represents $1/10^4$.

$$10^4 \times 10^2 = 10^{4+2} = 10^6$$
$$10^4 \times 10^{-3} = 10^{4-3} = 10^1 = 10$$

The fact that $10^0 = 1$ makes this multiplication, by adding and subtracting exponents, work in all cases.

Scientists use standard prefixes to modify **units**, such as meters (m) and make new units convenient for larger or smaller quantities. M, mega-, means 10^6 of the unit it modifies: 1 Mm = 10^6 m. 1 km = 10^3 m (kilo-); 1 mm = 10^{-3} (milli-); 1 micron = 10^{-6} m (micro-); and 1 nm = 10^{-9} (nano-). There are others, but these are the ones used often in these Keys. They can be used to modify units of time, speed, charge, and so on.

INDIVIDUAL KEYS IN THIS THEME	
1	Speed, velocity, and acceleration
2	Free fall
3	Motion in a horizontal plane
4	Motion in a vertical plane
5	Key questions with answers

Key 1   Speed, velocity, and acceleration

OVERVIEW *In your automobile, the speedometer tells you how fast you are moving over the earth's surface at the moment when you read it. It does not tell your direction of travel. Your average speed for a trip and your speed at one moment may differ. Velocity, a vector, has magnitude equal to speed and direction equal to the direction of motion. Velocity is the rate of change of position. Acceleration, also a vector, is the rate of change of velocity.*

Distance and position: The distance between two points can be described by saying how many units of distance lie on a straight line between the points.

- Two homes may be three miles (mi) apart; two persons 1.5 meters (m) apart, or two letters on this page four centimeters (cm) apart.
- Distances do not carry information about the direction from one point to another; they are *scalar* quantities.
- "The student union is 200 m west of this building" tells a distance and a direction—a displacement—the statement describes a *vector* quantity. Vector positions always refer to some origin, whether stated as in the example just given or implied.
- City addresses like 1476 North Fourteenth Street tell how far north and how far east or west a building is from the center of the city.

Speed and velocity: If an object moves in a straight line over the surface of the earth, its distance (*d*) from the starting point increases as time (*t*) passes. The rate of change of this distance is the *speed* (*v*) of the object:

$$v = \frac{d_2 - d_1}{t_2 - t_1}$$

The object's *velocity* **v** is its speed combined with the direction in which it is moving: **v** = 10 m/s, east.

- You can drive an automobile for two hours (h) at speeds that vary between zero miles per hour (mph) and 65 mph. Your average speed for the 2 h is the difference between the odometer readings at the start and at the finish divided by the time of the trip:

$$v_{avg} = \frac{15,728 \; mi - 15,628 \; mi}{2 \; h} = 50 \; mph$$

Your instantaneous speed at any time is what the speedometer shows then.

- Your instantaneous velocity at any time is your speed combined with the direction in which you are driving then: 35 mph northwest for example. Your average velocity is your displacement in the 2 h period divided by the time. Displacement is a vector. If, at the end of the trip, your automobile is 20 mi south of its starting point, then your average velocity was $\mathbf{v} = 10$ mph, south.

- If you had made a round trip in the 2 h, returning to your starting point, then your average velocity was zero because, in the end, your displacement was zero. However, your average speed and the distance your automobile traveled were both greater than zero.

Acceleration: Just as velocity is the rate of change of position, both vectors, acceleration $\mathbf{a}$ is the rate of change of velocity, also a vector:

$$\mathbf{a} = \frac{\mathbf{v}_2 - \mathbf{v}_1}{t_2 - t_1}$$

- Acceleration may have the same direction as velocity, for example, when you step on the gas to increase your automobile's speed on a straight road. When you brake your moving car to reduce its speed on a straight road, your velocity points ahead but your acceleration has the opposite direction—to your rear.

- Acceleration perpendicular to an object's velocity produces a change in the direction of the velocity, but not in the speed. The magnitude of the acceleration toward the center of the circle of an object moving along it at constant speed v is

$$a = \frac{v^2}{r}$$

where r is the radius of the circular trajectory.

OVERVIEW *Massive objects left unsupported near the earth's surface accelerate downward at a rate of approximately 10 m/s². This Key describes how the displacement and velocity of such freely falling objects evolve.*

The acceleration produced by gravity: The acceleration experienced by all unsupported objects near the earth's surface is the acceleration produced by gravity:

$$\mathbf{g} = 9.80 \text{ m/s}^2\text{, down}$$

It is often convenient and sufficiently accurate to use 10 m/s² for the magnitude *g*.

- To describe a free fall that starts from rest (zero velocity), take the moment when the fall starts as time $t = 0$ and measure position using distance down from the starting location. The instantaneous speed at time t is simply $v = gt$. The distance fallen at the end of time t is t times the average speed during the time interval. In motion with constant acceleration, such as free fall, the average velocity in any time interval is the **simple average** of the velocity at the start, $\mathbf{v}_1$, and the velocity at the end, $\mathbf{v}_2$:

$$\mathbf{v}_{avg} = \frac{\mathbf{v}_2 + \mathbf{v}_1}{2}$$

- In the free fall from rest, $v_1 = 0$ and $v_2 = gt$, so v_{avg} in the time interval from 0 to t is

$$v_{avg} = \frac{1}{2}gt$$

so that the distance fallen after time t is

$$d = \frac{1}{2}gt^2$$

- It is easy (very easy if you use $g = 10$ m/s²) to make a table with columns showing time, velocity, and displacement from the starting point at times of 1 s, 2 s, . . . , 10 s or more. You should make such a table: 1 s, 10 m/s, 5 m; 2 s, 20 m/s, 20 m; . . .

OVERVIEW *Motion in one direction or in two perpendicular directions, such as east-west and north-south, can be treated separately and then combined, or **superposed**, to describe the path, or **trajectory**, of a more complicated motion.*

Superposition in one dimension:
- When you swim upstream at a constant speed in a river, you move through the water faster than you move past the trees on the riverbank. For example, if you swim upstream at 1 m/s against a current flowing at 0.4 m/s, you move past the bank at 0.6 m/s. Do this for 100 s. In that time you move 100 m upstream through the water, while all the water around you moves downstream 40 m past the banks. Your net displacement is the vector sum of these two, 60 m upstream.
- Note that in one dimension you can completely specify vectors by using signed numbers. Positive displacements or velocities point one way; negative ones point in the opposite direction.
- Swim downstream at the same pace in the same river, and your displacement or velocity relative to the banks is greater than relative to the water. In 100 s you move 100 m downstream, while the water around you moves 40 m down; you move 140 m downstream relative to the banks.

Superposition in two horizontal dimensions:
- Let us continue swimming in the same river, supposing that it flows south at 0.4 m/s. Swim across the current, toward the west, at 1 m/s. If the river is 50 m wide, it takes you 50 s to get across. But you will not land at the spot you aimed at when you started. While you were swimming, the water moved south 20 m. Your displacement in the 50 s was along a straight line south of due west. It is just the same as if you had moved first 50 m west, then 2 m south. The Pythagorean theorem gives the distance you covered over the river's bed:

$$d = \sqrt{(d^2_{\text{west}} + d^2_{\text{south}})} = \sqrt{(2900) \text{ m}} = 53.9 \text{ m}$$

The Pythagorean theorem can also be used to compute your speed relative to the river bed:

$$v = \sqrt{(v^2_{west} + v^2_{south})} = \sqrt{(1.16)} \text{ m/s} = 1.08 \text{ m/s}$$

- You can treat any motion at constant velocity in a horizontal plane as two motions occurring simultaneously—one east-west and one north-south. Any information about the progress of the two-dimensional motion that you need can be obtained by calculating the results of two one-dimensional motions separately and then superposing them.

OVERVIEW *In the previous Key, we considered two-dimensional motion with constant velocity. A thrown ball moves in two dimensions with constant vertical acceleration and thus ever-changing vertical velocity. This Key describes that motion.*

Rest, constant velocity, and free fall:

- An object supported 20 m below the top of a tower will remain there at rest. Its position at all times is +20 m (measuring down from the top). An object constrained to move down at a constant speed of 30 m/s, starting at the top at $t = 0$, has position $+(30 \text{ m/s})t$. An object falling freely starting at rest at the top at $t = 0$ has position $\frac{1}{2}gt^2$. Superposing these three different kinds of motion gives the position of an object that is 20 m below the top at $t = 0$ with a velocity of +30 m/s (down—note that the positive numbers represent vectors pointing down; negative numbers represent vectors pointing up). After $t = 0$, the velocity of the object changes at a rate of **g,** +9.80 m/s^2 down. Its position at any time t is

$$d = 20 \text{ m} + (30 \text{ m/s})t + \frac{1}{2}gt^2$$

  Other starting positions and velocities, negative (up in this case) as well as positive, could be treated in the same way.

- To describe a ball thrown straight up in the air, one must use some vectors that point up, others that point down. Let us do so, changing first to a coordinate system in which positive vectors point up and position zero is ground level. For simplicity we shall approximate **g** by −10 m/s^2, negative meaning down.

$$d = d_0 + v_0 t + \frac{1}{2}(-10 \text{ m/s}^2)t^2$$

  For a ball thrown up from an initial height of 2 m, with an initial speed of 20 m/s, $d_0 = +2$ m and $v_0 = +20$ m/s. The motion is described by the superposition of: sitting at rest at a height of 2 m; moving up from ground level with constant speed of 20 m/s; and falling from rest at height 0 with constant acceleration **g**.

- The acceleration (negative or down) decreases the magnitude of the velocity (positive or up) after the motion starts. At what time, t_{max}, does the ball reach its greatest height, d_{max}, and what is that height? At what time, t_{end}, has the ball fallen back to a height of just 2 m, and what is the ball's velocity then? To begin to answer questions like these, consider how the velocity changes. Although positive at the start and decreasing, it must eventually pass through zero and take on negative values. The time when the velocity is zero is the time when the ball is at its highest point; before that time the ball was moving up, and after it the ball is moving down.

- The initial velocity decreases by $g \times 1$ s in each second after the start, so t_{max} is just the time such that $gt_{max} = v_0$ or

$$t_{max} = v_0/g$$

- The average velocity in the time interval from 0 to t_{max} is just the simple average of v_0 and 0, $\frac{1}{2}v_0$. Using this, you can find how high above d_0 the ball was at t_{max}:

$$d_{max} - d_0 = \frac{1}{2}v_0 t_{max} = \frac{1}{2}gt^2_{max}$$

The thrown ball rises in time t_{max} to a height $d_{max} - d_0$ above its starting point that is the same as the distance it would fall from rest in a time t_{max}. Using this fact, the time when the ball passes its starting height d_0 on the way back down is $2t_{max}$; it takes t_{max} to rise to the peak and another t_{max} to fall from rest there back to its starting height. Falling from rest at its peak height during the second interval of duration t_{max}, the ball acquires velocity

$$gt_{max} = -(10 \text{ m/s}^2)t_{max} = -v_0$$

$-g \times 1$ s during each second of the fall. The ball passes its starting point moving down at the same speed with which it moved up through that point at the start, if air friction can be neglected.

Projectile motion: To describe the motion of a cannonball or a baseball projected upward and outward, rather than straight up, requires very little more work, if one is clever enough to realize that motion in a vertical plane can be treated by **superposing** the vertical motion just described and horizontal motion at constant velocity.

- Consider the ball of the previous section, thrown with an initial velocity of v_0 up and v_1 south. It reaches its greatest height, d_{max}, at time t_{max}, and returns to its starting height at $2t_{max}$. The point on the ground directly under the position of the ball at its greatest height is $v_1 t_{max}$ south of the starting point. The ball returns to its original height, d_0, at a distance $2v_1 t_{max}$ south of its starting point.
- If you happened to throw the ball from the north rim of the Grand Canyon, then it would continue, long after $2t_{max}$, to accelerate down at a rate g and move south at a constant speed v_1.

Very fast projectiles—satellites: Man-made earth satellites just above our atmosphere move with speeds around 7.9×10^3 m/s, or 16,000 mph. They fall around earth in their near-circular orbits. In 1 s, one drops 5 m but moves east (usually) so far that the curvature of the earth's surface has made the surface "drop" 5 m also.

OVERVIEW *Sample questions of the type that might appear on homework assignments and tests are presented with answers.*

Distance, speed, and time:
- An astronaut sits in her super-shuttle 7×10^8 m above the center of Jupiter. The shuttle moves in a circular orbit with speed 1.4×10^4 m/s. What is the period of its orbit—the time to go around the planet just once? It is $T = 2\pi R/v$ or $2\pi(7 \times 10^8$ m)/$(10^4$ m/s) which is 4.4×10^5 s.

Acceleration:
- In one second, a car moving straight ahead increases its speed from zero kilometers per hour (km/h) to 10 km/h. In 0.1 s, a rocket moving straight up increases its speed from 1000.0 km/h to 1000.5 km/h. Which had the greater acceleration? Car: 10 km/(hs). Rocket: (0.5/0.1) km/(hs) = 5 km/(hs). The car had the greater acceleration.

Free fall near the earth's surface:
- A pitcher throws a baseball with an initial velocity of 33 m/s, straight down from a cliff. Neglecting air friction, what is its acceleration just after it leaves the pitcher's hand? It is $g = 10$ m/s^{-2}, down.
- If you could open a window in an office building in Chicago and drop a rock, you might observe that it takes 8.0 s for it to reach the ground. How high above the ground are you? $d = \frac{1}{2}gt^2 = 5$ ms$^{-2}(8.0$ s$)^2 = 320$ m.

Motion in a horizontal plane:
- A pilot points his boat straight east and runs for 2 h at 5 mph relative to the water. There is a current running from the south (toward the north) at 1 mph. How far east does the boat move, relative to the land? How far north does the boat move, relative to the land? East: 5 mph × 2 h = 10 mi. North: 1 mph × 2 h = 2 mi.

Projectile motion:

- A cannonball is shot in a direction above the horizontal. Its starting velocity is 200 m/s in the horizontal direction and 50 m/s in the vertical direction. Neglecting air friction, how long does it take the ball to reach its highest point? How high above the ground is it at its highest point? How far in the horizontal direction from the cannon is the point on the ground where the ball lands? The 50 m/s initial vertical speed divided by $g = 10$ m/s = 5 s time until the vertical velocity is zero at the peak of the trajectory. The 25 m/s average vertical velocity during the first 5 s produces an upward displacement of 25 m/s × 5 s = 125 m at the highest point. The 200 m/s horizontal speed × 5 s = 1000 m distance from the cannon to the point on the ground beneath the highest point. On its way down, the ball travels this far again, landing 2000 m from the cannon.

Theme 2   NEWTON'S LAWS OF MOTION

This Theme presents the precise relationships among the complex concepts of acceleration, force, and inertia. Galileo worked with each of the concepts, using Euclid's geometry as his analytical tool, but did not find the laws that Newton stated later. With Newton's laws and algebra as his or her analytical tool, any college student can understand the concepts and their relationships.

Key 6 Newton's first law of motion, inertia, and mass

OVERVIEW *A moving object resists any change in its velocity. This property of all objects is called inertia. Mass is the quantitative measure of an object's inertia.*

The motion of an isolated object: Newton's first law states that the velocity of an isolated object does not change as time passes. But what does "isolated" mean?

- A rocket ship coasting between the earth and the moon, with its engines off, behaves for minutes or hours as if it were completely isolated from its environment.
- A heavy box of books sliding down a concrete loading ramp does not behave as if it were isolated, not even for one second. There is an interaction between the box and the concrete ramp where they are in contact, **friction,** that makes the velocity of the box change. The sliding box stops.

Inertia: Find a golf ball and a ping-pong ball. Cup one between your hands and move them back and forth rapidly. You force the ball to change its velocity frequently. You can tell which ball is in your hands without looking, because you must push hard to change the golf ball's velocity, but less hard to change the ping-pong ball's. The golf ball has greater inertia; it more effectively resists any change in its velocity than does the ping-pong ball.

Mass: Mass is a quantitative measure of inertia. You may think of it as measuring how much matter there is in an object. Mass has simple mathematical properties that are required for a **useful** measure of inertia.

- If you break an object into fragments, the sum of the masses of the fragments is equal to the mass of the original object.
- The mass of two golf balls is twice the mass of one alone. The same is true for two ping-pong balls. The mass of a golf ball and a ping-pong ball together is only a little more than the mass of a golf ball alone.

Units of mass: The standard (SI) unit of mass is the kilogram (kg).

- A cubic meter of water has a mass of 1000 kg.
- A cubic centimeter of water has a mass of 1 g.
- Different persons have different masses, but 70 kg is typical.
- A mug full of coffee has a mass of about 0.4 kg.

Newton's second law of motion, force

OVERVIEW *When you push on something, force is the quantitative measure of how hard you push. You must push hard to change the velocity of an object with great inertia, less hard to accelerate an object with less inertia.*

Force: Force, like mass, has simple mathematical properties that make it a **useful** measure of how hard something pushes.
- Two identical compressed springs pressing on an object side by side produce twice as much force on the object as one of them would alone. These are contact forces because they act only when a spring touches an object.
- Like velocity and acceleration, force is a vector quantity having both a magnitude and a direction. (How hard are you pushing and which way are you pushing?)

Units of force: The SI unit of force is the Newton (N). The English unit of force is the pound (lb). One newton is about 0.22 lbs.
- A full mug of coffee weighs about 1 lb or 4.45 N.
- A typical person weighs about 150 lbs or 670 N.

Motion of a nonisolated object: Newton's second law states that the acceleration of an object that has outside forces acting on it is directly proportional to the **net** force and in the direction of the **net** force, and that the acceleration is inversely proportional to the object's mass:

$$a = \frac{F}{m}$$

- The net force is the vector sum of all the individual forces acting on an object. If you see an object accelerating, you know that the net force on it is not zero; if its velocity is constant, on the other hand, you know that the net force on it is zero.
- An object is in **mechanical equilibrium** if it is not accelerating, either because no force acts on it or because the forces that are acting on it add (vectorially) to zero.

Units of force, again: One newton is the magnitude of force that will cause an object with a mass of 1 kg to accelerate at a rate of 1 m/s^2. Because **F** = m**a**, 1 N is 1 kg • m/s^2.

Key 8 Free fall and nonfree fall

OVERVIEW *Because gravity produces acceleration,* **g**, *there must be a force of gravity. An object that is falling while acted upon by this force alone is in free fall. If any other force acts, then the fall is not free.*

The force produced by gravity:
- All objects in free fall near the earth's surface have the same acceleration, $g = 9.8$ m/s^2, down. The force of gravity that produces the acceleration must be $F = m\mathbf{g}$. This is a noncontact or action-at-a-distance force.
- To produce the same acceleration for every falling object, the gravitational force must be proportional to m—the mass of the object.

Free fall:
- On the earth's surface, we are aware of weight, the force of gravity acting on us when we are standing at rest, because we feel the other force, pushing on our feet, that just cancels the gravitational force. In free fall we would feel no force.
- Astronauts in the space shuttle, in orbit around the earth with the engines off, are in free fall. So is the shuttle. Because shuttle and astronauts accelerate at the same rate, the astronauts perceive no acceleration of themselves relative to the shuttle. They can float inside it for a long time without "falling" to one of the walls. The astronauts feel like no force is acting on them, because gravity gives each part of their bodies the same acceleration. It does not produce any tension or compression in their muscles that they could feel.

Nonfree fall:
- A feather that falls from 2 m above the earth's surface is not in free fall. It accelerates briefly at the start; then falls the rest of the way with approximately constant velocity. Air friction produces a force on it that opposes the gravitational force. The force of air friction increases as the speed of an object through the air increases.
- Falling people use parachutes to increase air friction a lot without increasing their mass very much. This reduces their terminal velocity, the constant velocity at which the force of air friction just cancels that of gravity, to a nonharmful value.

OVERVIEW *If an object* a *exerts a force on another object* b, *then* b *must at the same moment exert an equal and opposite reaction force on* a. *This is Newton's third law of motion.*

Experience a third-law reaction force: Stand still facing a wall while wearing roller skates. Push on the wall with your hands. A moment later you find yourself rolling away from it. You accelerated; the wall must have exerted a force on you.

Reaction forces in transportation: Automobiles, power boats, airplanes, and rockets are all propelled by reaction forces. Tires exert forces against roads, propellers against water or air, rocket motors against the gases they expel. In each case, the road, water, air, or exhaust gas pushes back against the vehicle, either balancing frictional forces to maintain constant velocity or exceeding them to produce acceleration.

Recoil and relative mass:
- When a bullet with a mass of 1 g is shot from a rifle with a mass of 500 g, the force on the rifle has the same magnitude as that on the bullet. The bullet accelerates to high velocity, say 280 m/s, in a time of about 4 ms, with average acceleration 7×10^4 m/s^2 during that time. The rifle also accelerates, or recoils, back against your shoulder. Its acceleration is 1.4×10^2 m/s^2, which is smaller than the bullet's because its mass is greater. In the 4 ms interval it moves back 1.4 cm and acquires a velocity of 0.6 m/s; in the next 100 ms or so your shoulder stops it easily, without damage (the average force on your shoulder is 12 N or 2.7 lbs).
- Consider an ultralight, high-power rifle designed to shoot a heavy bullet, so that rifle and bullet have equal masses. You would damage your shoulder if you attempted to shoot it.

Make the earth move:
- Stand on a chair. Step off and fall 0.5 m to the floor. You accelerate downward, move down 0.5 m, because of the force of the gravity produced by the earth, acting on you, during your fall. You produce a force that acts on the earth during a fall. The earth accelerates upward and moves up a little bit while you fall. Its motion is imperceptible because its mass is so much greater than yours.

OVERVIEW *Sample questions of the type that might appear on homework assignments and tests are presented with answers.*

The second law:

- A force of 1 N accelerates a mass of 1 kg at a rate of 1 ms^{-2}. A force of 1 N accelerates a mass of 10 kg at what rate? The acceleration is $a = F/m$ or 1 N/10 kg, which is 0.1 ms^{-2}.

- If you hold a pizza that has a mass 1.0 kg still the force OF GRAVITY on it is how many N, in what direction? The force is $F = mg = 1.0$ kg $\times$ 10 ms^{-2} = 10 N down.

- If you hold a pizza that has a mass of 1.0 kg still, what is the NET force on the pizza (i.e., the sum of all forces acting on it), taking account of directions? The net force is zero if the acceleration is zero.

- An object with mass 47 kg accelerates at 19 ms^{-2}. What is the net force acting on it? $F = ma$, which is 47 kg $\times$ 19 ms^{-2} or 8.9×10^2 N.

- An object with mass 47 kg is acted on by two forces: 17 N north and 27 N west. What is its acceleration? It is easiest to give the components of the acceleration in the directions north and west. $a = F/m$, so the acceleration to the north is 47 kg/17 N or 2.8 ms^{-2} and that to the west is 47 kg/27 N or 1.7 ms^{-2}.

Air friction:

- A pitcher throws a baseball with an initial velocity of 33 m/s, straight down from a cliff. Air friction tends to reduce the speed of the accelerating ball. In what direction is the force of air friction just after the ball leaves the pitcher's hand? The force of air friction is in a direction opposite to the ball's velocity, up in this case.

MOMENTUM, WORK, AND ENERGY

Momentum and energy are two different measures of how hard it would be to stop a moving mass, and how hard it was to put it into motion. Galileo, unaware of the difference between these concepts, enjoyed only a limited use of them. Huygens, Newton, and Leibniz defined momentum and energy precisely in terms of time, distance, velocity, mass, and force. Now, any college student can use the definitions to answer in a few lines mechanics questions that Galileo needed pages to answer.

Key 11 Momentum and impulse

OVERVIEW *The momentum of an object is defined as the product of its mass, a scalar, times its velocity, a vector. Momentum is a vector with the same direction as that of the velocity and with a magnitude that is the product of the mass and the speed of the object. The impulse given to an object is the product of the time interval during which a force acts on the object, a scalar, and the average force, a vector, acting during the interval. Impulse is a vector with the same direction as that of the force. The change in the momentum of an object during a particular time interval is equal to the impulse it receives during that interval.*

Momentum: An object can have a large momentum, $\mathbf{p} = m\mathbf{v}$, if at least one of m and $\mathbf{v}$ has a large magnitude.
- A bullet with mass 0.01 kg and speed 200 m/s (about 450 mph) has the same momentum as a 1000 kg automobile with speed 2×10^{-3} m/s (2 mm/s).
- To change the momentum of an object with constant mass, you must change its velocity (i.e., accelerate it). To do so you must apply a force.

Impulse: An impulse $\mathbf{F}t$ can have a large magnitude if at least one of $\mathbf{F}$ and t is large.
- A force of 2000 N (about 450 lbs) acting for 10^{-3} s produces the same impulse as a force of 2 N (about 0.45 lbs) acting for 1 s; the same as 2×10^{-3} N acting for 1000 s.

Newton's second law: In the simple case of constant acceleration from rest in response to an applied force that is constant for a time interval $t_f - t_i$

$$\mathbf{a} = \frac{\mathbf{F}}{m}$$

$$\mathbf{v} = \mathbf{a} \, (t_f - t_i) = \frac{\mathbf{F}(t_f - t_i)}{m}$$

$$\mathbf{p} = m\mathbf{v} = \mathbf{F} \, (t_f - t_i)$$

where the object accelerated has mass m and acquires velocity $\mathbf{v}$ at the end of the interval.

19

- More generally, but as a direct consequence of the second law,

$$\mathbf{p}_f - \mathbf{p}_i = m\mathbf{v}_f - m\mathbf{v}_i = \mathbf{F}\,(t_f - t_i)$$

where the subscripts i and f label quantities before and after the impulse was given.
- An impulse produces a change in momentum that is equal to the impulse in magnitude and in direction:

$$\frac{\mathbf{p}_f - \mathbf{p}_i}{t_f - t_i} = \frac{m\mathbf{v}_f - m\mathbf{v}_i}{t_f - t_i} = \mathbf{F}$$

This last form of the second law shows that, just as acceleration is the rate of change of velocity, net force is the rate of change of momentum.

Producing and absorbing momentum: You can either produce or absorb the momentum of a moving object by applying a large force for a short time, or a smaller force for a longer time.
- Both the automobile and the bullet in the first section of this Key have momenta with magnitude 2 kg • m/s. Each of the impulses described in the second section of this Key has just this magnitude—enough to change the momentum of the car or of the bullet from zero to 2 kg • m/s, or from 2 kg • m/s to zero.
- Passenger cars have collapsible steering columns and padded dashboards to extend the time during which occupants lose their momenta in a crash, and so further reduce the forces they suffer.
- Athletes in contact sports actively control time intervals during which momenta are absorbed by their bodies, sometimes to minimize forces and at other times to maximize them. Football and boxing each provide examples of both kinds of control. Can you discover them for yourself by watching carefully?

Units of momentum and impulse: The standard (SI) unit of momentum is 1 kg • m/s.
- An impulse of 1 Ns (a force of 1 N acting for a time of 1 s, for example) produces a 1 kg • m/s change in momentum. 1 N = 1 kg • m/s^2 (see the last section of Key 7) so 1 Ns = 1 kg • m/s, as required by the form of the second law given in the third section of this Key.
- A pitched baseball has a momentum of about 10 kg • m/s.
- An automobile on the highway has a momentum of about 3×10^4 kg • m/s.
- A person walking briskly has momentum of about 400 kg • m/s.

Key 12 Conservation of momentum and collisions

OVERVIEW *The total (vector sum) momentum of a system of massive objects changes only if an outside force acts on the system. Internal forces between the objects can redistribute the total momentum among them but cannot change the total. If the external force is zero, the total momentum of the system is conserved—it remains constant.*

Newton's second and third laws: It follows directly from the second law as written in the third section of Key 11 and the third law (Overview of Key 9) that two massive objects exerting forces on each other experience, as a result, equal and opposite changes in their momenta.

- The vector sum of their two momenta remains constant as long as no outside force acts on them.
- Equally directly, although with more complicated-looking formulae, it follows that the total momentum **P** of any number N of interacting massive objects is conserved if no external force acts on them, and that the change in the total momentum in a time interval $t_f - t_i$ is equal to the total (vector sum) impulse that the system receives from outside during the interval:

$$\mathbf{P} = \mathbf{p}_1 + \mathbf{p}_2 + \dots + \mathbf{p}_N$$

$$\mathbf{F}_{ext} = \mathbf{F}_{ext\ on\ 1} + \mathbf{F}_{ext\ on\ 2} + \dots + \mathbf{F}_{ext\ on\ N}$$

$$\mathbf{P}_f - \mathbf{P}_i = \mathbf{F}_{ext}\ (t_f - t_i)$$

$$\frac{\mathbf{p}_f - \mathbf{p}_i}{t_f - t_i} = \mathbf{F}_{ext}$$

Collisions: Before, during, and after a collision between two or more massive objects that move free from friction or other external forces, the sum of their momenta is constant. This is true whether the objects collide and rebound with no permanent change of shape or generation of heat, like billiards balls or air-track gliders equipped with spring bumpers, or whether there is permanent deformation or even sticking together, like automobiles that lock their bumpers together in a crash on an icy road.

Simple examples in one direction: Consider two objects, each having mass 4 kg. Let the one on your right be at rest before the collision, and the one on the left have velocity 5 m/s to the right initially. The total momentum of the system is the vector sum of zero (right object) and 20 kg • m/s to the right (left object). Imagine that there is no friction, no external force.

1. If the objects are stuck together after the collision, then they move as a single mass of 8 kg. To have the same total momentum as before, their common velocity must be

$$\frac{20 \text{ kg} \bullet \text{m/s}}{8 \text{ kg}} = 2.5 \text{ m/s}$$

2. If the objects rebound without sticking together, there are many possibilities. The one on the left might stop and the one on the right move to the right at 5 m/s; the total momentum in this state of motion is the same as in the initial state. If the one on the right moves to the right at 8 m/s, and the one on the left moves to the left at 3 m/s, the total momentum in this state is also the same. One needs more information to determine the final state. If the collision occurred without generating heat or permanent deformation, the final state is the first described, in which the object on the left is at rest. Key 15 describes the additional fact, also derived from Newton's laws, that determines this unique solution.

Collisions in two and three dimensions: In each of two or three orthogonal directions, such as up-down, north-south, or east-west, the component of total momentum remains constant if the external force has zero component along the selected direction, or the component of momentum changes in a time interval t by an amount equal to the component of the external impulse along the selected direction that was received during the interval. Two- and three-dimensional collisions can be analyzed in the same way as one-dimensional collisions.

Key 13 Work and power

OVERVIEW   *Just as impulse produces a change in momentum, work produces a change in energy, the second of the two quantities that Galileo thought of as one. Work, like energy, is a scalar, in contrast to vector momentum and impulse. Power is the rate of doing work.*

Work: The work that an agent does on an object is defined as the scalar product of two vector quantities, the force applied by the agent to the object and the displacement of the object while the force is present.

- Work W is computed as the product of the magnitude of the force and the component of the displacement parallel to the force, as the product of the magnitude of the displacement and the component of the force parallel to the displacement, or as the product of the two magnitudes and the cosine of the angle θ between the two vectors:

$$W = Fd_{\text{parallel}}$$
$$W = F_{\text{parallel}}d$$
$$W = Fd\cos(\theta)$$

- Work done by an agent on an object is positive if the agent's force and the component of the displacement parallel to it point in the same direction; if their directions are opposite the work is negative.
- Because of the third law, whenever an agent does work W on an object, the object does work $-W$ on the agent.
- A force perpendicular to the direction of an object's motion produces no work.
- A force on an object that remains at rest produces no work.

Power: Power P is the rate at which work is done. If an agent does work W in a time interval of duration t, the rate at which an agent does work is

$$P = \frac{W}{t}$$

SI units of work and power: A force of 1 N applied in the direction of motion of an object while it moves 1 m produces work $W = 1$ N • m. This amount of work is called one joule, 1 J after James Joule, a nineteenth-century British physicist. If 1 J of work is done in a time of 1 s, the power 1 J/s = 1 W. The watt was named after James Watt, an early nineteenth-century British engineer.

- You do about 9 J of work while lifting a cup of coffee from the table to your lips. If you complete the lift in 1.5 s your power was 6 W.

- An automobile engine that produces 150 horsepower (150 hp) delivers 112 kW power; 1 hp is about 0.75 kW.

- Electrical energy delivered to a home is commonly measured in kilowatt-hours. kWh. 1 kWh = 1 kW $\times$ 3600 s = 3.6 MJ. To keep a 100 W bulb lighted one hour, some agent must do .36 MJ work, at a rate of 100 W.

OVERVIEW *In some situations work done by an agent may be stored away, recoverable at a later time. For example, you could stand on a playground and kick a soccer ball to make it speed across the ground right now. You might instead lift the ball up and place it at the top of a children's slide. It would stay there as long as you wish, to speed across the ground whenever you start it with a gentle nudge. The speed it acquires is the result of the work you did earlier in lifting it up. Work stored away is potential energy. The energy a massive object has because of its speed is kinetic energy.*

Mechanical potential energy: Drawing a bowstring back, compressing a spring, and inflating a balloon are some of the ways in which you can store mechanical energy. Changing a battery produces nonmechanical potential energy, chemical potential energy.

- The mechanical potential energy you give you an object while lifting it is equal to the work that you do in the lift. To move the object upward at a low speed, you must apply a force equal and opposite to the object's weight mg.

- Raising an object to a height h above its starting position increases its potential energy by mgh. The path by which you lift it does not matter; straight up, up an incline, or up by any other path requires the same work and results in the same potential energy, because of the scalar-product relation between work, force, and displacement (first section of Key 13).

- There is an undefined, constant amount of energy in the gravitational potential energy of any object near the earth's surface, which depends on the height h_0 at which you call the potential energy zero.

- In solving a problem, choose any convenient height to be h_0, and do not change it until the problem is finished. The change in an object's potential energy as it moves from h_1 to h_2 is $mg(h_2 - h_1)$, which is independent of your choice of h_0.

Kinetic energy: Kinetic energy is the energy associated with the motion of a massive object. Kinetic energy is easily computed by calculating the work required to accelerate the object at a constant rate a from rest to speed v in a time t during which the object moves through a distance d:

$$W = Fd$$

$$W = ma \frac{1}{2} at^2$$

$$W = \frac{1}{2} m(at)^2$$

$$W = \frac{1}{2} mv^2$$

- The kinetic energy depends only on the mass and speed of the object. It can be shown that it is independent of how the object attained its speed.

Units of energy and examples: Energy has the same units as work, force times distance. One newton meter equals one joule, the unit described in the third section of Key 13. Kinetic energy has dimensions mass times speed squared, but the derivation in the preceding section proves that $1 \text{ kg(m/s)}^2 = 1 \text{ J}$.

- A soccer ball on top of a playground slide, 2 m above the ground, has about 8 J more potential energy that it would have on the ground. If it rolls down the slide without friction, all this is converted to kinetic energy, resulting in a speed of about 6.3 m/s.
- Your kinetic energy while walking is about 35 J. Climb up on a 1 m high stool and you increase your potential energy by about 700 J; if you jump off, your speed just before you land on the floor will be almost five times walking speed.
- A high diver on the 10 m platform has almost 7×10^3 J more potential energy than on the pool deck. If air friction can be neglected, then the diver's speed just before entering the water is 14 m/s, about three times that acquired by diving off a 1 m board.
- An automobile with mass 1.3×10^3 kg on an interstate highway has about 3.9×10^5 J kinetic energy when it moves at the 55 mph (24.6 m/s) speed limit. To give it this much potential energy, you would have to raise it 31 m.

Conservation of energy
and efficiency

OVERVIEW *Mechanical energy remains constant or approximately constant in some situations. If a system is not subject to external forces, and if none of the forces within the system arise from friction, then the system does not produce heat and none of its parts suffer permanent deformation as time passes. In this case the energy of the system is conserved; it does not change. If in addition the system's mechanical energy is not changed into other forms, such as chemical energy, then its mechanical energy—kinetic plus potential—is conserved. A system with friction loses mechanical energy as time goes on and heat is produced. When energy put in and energy obtained back can be identified and measured, the system's efficiency is defined as the ratio of energy obtained to energy put in. Rather than give up the concept of energy conservation in situations with friction, physicists have succeeded in treating heat as another form of energy and using the principle that the sum of all forms of energy in an isolated system is conserved.*

Examples:
- Consider a diver with a mass 70 kg on a 10 m diving platform. Her potential energy is almost 7000 J greater than at the water's surface. Standing still, her kinetic energy is zero. In a dive, 3 m below the platform, her potential energy is about 2100 J smaller, and her kinetic energy is about 2100 J. Neglecting air friction, her total mechanical energy is the same as before she dove. Her speed at this point is

$$v = \sqrt{\frac{2KE}{m}}$$

where *KE* is her kinetic energy and *m* is her mass. Her speed is almost 8 m/s. A bit later she is 6 m below the platform and her potential energy is about 4200 J less than before she dove. Her kinetic energy is equal to this amount because, again neglecting air friction, her total mechanical energy remains constant. Her speed here is almost 11 m/s. Just before entering the water, all the potential energy she gained by climbing the tower has converted to kinetic energy, almost 7000 J and her speed is about 14 m/s.

- A pendulum swings with its speed and its height above its rest position varying so that, neglecting friction from the air and the support, its total energy

$$mg(h - h_0) + \frac{1}{2}mv^2$$

remains constant. Its energy is all potential as it momentarily stops at its highest point and all kinetic when it speeds through its lowest point (h_0).

Almost conservative systems:
- A real pendulum, if carefully constructed, may swing a hundred times before it loses half its energy because of frictional forces from the air and in its bearing. Without oil in the bearing, the number of swings might be only two.
- A roller coaster with a vertical loop retains enough of its energy, which was mostly potential at the top of its first climb, to whiz around a loop upside down without leaving the tracks. But the hills over which it passes become lower and lower, not exciting at all at the end of the ride, when friction has stolen away most of its initial energy.

Efficiency: Efficiency is often defined as useful work obtained from a machine divided by the energy put into it.
- Charge up a battery and then use it to run a motor that produces mechanical work you consider useful—perhaps it lifts water from a well. Besides useful work, the system produces heat from the battery, from the wires between the battery and the motor, from the wires in the motor, from the motor bearings, and from other sources as well. This system is not 100 percent efficient. Useful work out plus heat out equals the energy put in when the battery was charged, but useful work out by itself is less than the energy put in.
- An automobile engine produces significant heat and noise, as well as work exerted against air friction, road friction, and friction within the machine. If you had a 100 percent efficient automobile, how much noise would it make?

OVERVIEW *Sample questions of the type that might appear on homework assignments and tests are presented with answers.*

Momentum and impulse:

- You launch a Rapid Mail Rocket (RMR-II) that has a mass of 1000 kg. It burns for 40 s, producing a NET force of 10^4 N on the rocket during that time. What is the momentum of the flying missile at the end of its burn? The final momentum is equal to the impulse $Ft = 10^4$ N $\times$ 40 s = 4×10^5 Ns = 4×10^5 kg • m/s.

- A Jesse James Mail Robber IV antiballistic missile weighs only 100 N. It has momentum 4×10^4 kg • m/s. What is its speed as it streaks toward the hapless RMR-II? Speed v is momentum divided by mass: $v = 4 \times 10^4$ kg • m/s/m, where m is the weight divided by g. $m = 10$ kg and $v = 4 \times 10^3$ m/s.

Work and power:

- A Chevrolet cruises at 30 m/s speed. To overcome air resistance and road friction, its engine must produce a push of 500 N. What is the power expended by the car against the air and road? The power is $P = Fv = 500$ N $\times$ 30 m/s = 1.5×10^4 W.

Kinetic and potential energy:

- You red Radio Flyer wagon with perfectly frictionless bearings coasts from a standing start at the top of a hill that is 20 m high. What is your speed when you reach the bottom of the hill? Potential energy mgh is converted to kinetic energy $\frac{1}{2}mv^2$. The m cancels when these are equated and $v = \sqrt{2gh} = 20$ m/s.

Conservation of momentum and collisions:

- A bowler standing still on ice, wearing ice skates, bowls the ball. The ball with a mass of 10 kg moves over the ice at 5 m/s north, after release. What, then, is the velocity of the bowler, who has mass 63 kg? The system has total momentum zero. After the release, the ball has momentum 10 kg $\times$ 5 m/s = 50 kg • m/s north. The bowler must have momentum 50 kg • m/s south and so must have velocity 50 kg • m/s / 63 kg = 0.8 m/s south.

- A small tugboat with mass 4×10^4 kg is moving north at 2 m/s. An open-class hydroplane with mass 2×10^3 kg is moving south with speed 40 m/s. They collide and stick together. If their momentum is conserved in the collision, what is their common velocity after the crash? Before the collision, the momentum of the tug is 2 m/s × (4 × 10^4 kg) north or 8 × 10^4 kg • m/s north. The momentum of the hydroplane is 40 m/s × (2 × 10^3 kg) south or 8 × 10^4 kg • m/s south. The total momentum is zero, so after the collision the stuck-together boats remain at rest.

Conservation of momentum, energy, and collisions:

- A small car, mass 10^3 kg, is moving east at 10 m/s, and a truck, mass 10^4 kg, is moving west at 2 m/s. What is the total momentum of the two vehicles? What is the total kinetic energy of the two vehicles before their collision? Which vehicle has the greater kinetic energy? The car's momentum is 10 m/s × 10^3 kg east, or 10^4 kg • m/s east, and its kinetic energy is $\frac{1}{2}$ × 10^3 kg × (10 m/s)^2 or 5 × 10^4 J. The truck's momentum is 2 m/s × 10^4 kg west, or 2 × 10^4 kg • m/s west, and its kinetic energy is $\frac{1}{2}$ × 10^4 kg × (2 m/s)^2 or 2 × 10^4 J. The total momentum is 10^4 kg • m/s west and the total kinetic energy is 7 × 10^4 J, the car having the greater kinetic energy.

- The car moves north at 20 m/s and the truck moves south at 2 m/s. If a perfectly elastic collision occurs, what is the velocity of the car after the collision? The car has momentum 2 × 10^4 kg • m/s north and the truck has momentum 2 × 10^4 kg • m/s south. The total is zero. Both the car and the truck can have the same kinetic energies after the collision as they had before and equal but opposite momenta, if their velocities simply reverse in the collision. The car moves south at 20 m/s after the collision.

- If, instead of the collision just described, the car and the truck suffer a maximally inelastic (sticking) collision, then what is the velocity of the car afterwards? The total momentum before the collision is zero, and so the velocity after a sticking collision must be zero.

- If, instead of the two collisions just described, a collision occurs from which the car emerges with a velocity of 10 m/s south, then how much mechanical energy has been lost (converted to nonmechanical forms of energy)? To conserve total momentum, the truck must move north at 1 m/s after the collision. Each vehicle has half its initial speed after the collision, and therefore one quarter its initial kinetic energy. Three fourths of the kinetic energy has been lost. The original kinetic energy of the car was $\frac{1}{2}$ × 1000 kg × (20 m/s)^2 or 2 × 10^5 J, and that of the truck was $\frac{1}{2}$ × 10000 kg × (2 m/s)^2 or 2 × 10^4 J, for a total of 2.2 × 10^5 J. The amount lost, three fourths of the original total, is 1.65 × 10^5 J.

Theme 4 ROTATIONAL MOTION

The concepts used to describe circular motion and motion along other curved trajectories are each analogous to one of the linear-motion concepts described in the first three Themes. The **analogy** between rotational and linear motion is used extensively in most textbooks; you must understand the preceding material on linear motion to learn about nonlinear motion. This Theme does not describe each analogous concept but does present the additional new concepts needed to treat rotational motion.

OVERVIEW *Just as a massive object resists any change in its velocity, so an extended, massive, rigid object resists any change in its rate of rotation or in the orientation of the axis of its rotation. The moment of inertia is the quantitative measure of this rotational inertia.*

The moment of inertia: A very small mass m at a distance r from an axis about which it moves in a circle has a moment of inertia (I).

$$I = mr^2$$

- The standard (SI) unit is 1 kg • m^2.
- The moment of inertia of an extended object has the same form: total mass multiplied by an average value of the square of the distance from the mass to the rotation axis.
- A ball of radius r and mass m has

$$I_{\text{ball}} = \frac{2}{5} mr^2$$

- A thin rod of length 1 and mass m has

$$I_{\text{rod}} = \frac{1}{12} ml^2$$

  about an axis through its center and perpendicular to its length.
- Grip your closed umbrella in the center and rotate it back and forth. The resistance you feel is measured by its moment of inertia, about 0.4 kg • m^2.

Mass distribution: Two objects with identical shape, size, and mass may resist changes in their rotation differently.
- Let two cylinders roll down a ramp starting from rest. Assume their lengths, masses, and diameters are identical, but one is a solid cylinder of a light material and the other a thin, cylindrical ring of a heavy material. The ring has all its mass far from its center, resists an increase in its angular speed more effectively, has the greater moment of inertia, and so reaches the bottom of the ramp later than the solid cylinder.

- Let us learn to walk a tightrope. Carry either a bowling ball or a 4 m long rod with the same mass. If you teeter a bit to the left, push down with your left hand and up with your right to recover. Which do you prefer to carry?

Key 18 Torque and angular momentum

OVERVIEW *The analogs of Newton's laws are the follow-ing. First, the angular acceleration of a rigid object not acted on by any net torque is zero. Second, the angular accelera-tion of an object is the net torque acting on it divided by its moment of inertia. Third, if object a exerts a torque on b, b exerts an equal and opposite torque on a. These analogs follow from Newton's laws for linear motion applied to the bits of mass in rigid extended objects. Angular momentum and its conservation follow from the same considerations.*

Torque: Torque is defined as the product of the part of a force F applied to an object that is perpendicular to its rotation axis, F′, times the **perpendicular** distance $d_{perpendicular}$ from the line along which the force acts to the rotation axis about which the torque is computed.

• Torque is frequently represented by Greek letter tau:

$$\tau = F'd_{perpendicular}$$

• A single force on an object generally produces different torques, if different rotation axes are considered.

• The direction of a torque is taken to be along the rotation axis; you can further describe it as tending to produce either clockwise or counterclockwise rotation about the axis. Your textbook or another general physics text describes a right-hand rule that relates the direc-tions of the rotation and the torque.

Angular momentum: The rotational analog of linear momentum is the angular momentum **L** of an object. Its magnitude is the product of the momentum of inertia (I) times the rate of rotation of the object.

• The angular momentum of an object subject to no net torque is conserved.

• If there is a net torque, the rate of change of angular momentum is equal to it. **L** may change in magnitude or in direction, possibly in a complicated manner, as in the case of a precessing gyroscope or top.

• When a torque is parallel to an angular momentum that points along the rotation axis, the angular momentum increases or decreases without changing direction.

Key 19 Center of mass, its motion, and stability

OVERVIEW *The center of mass of an object is the single point that moves as if all the mass were concentrated at that point. On a flat part of the earth's surface, an object resting on its flat base is stable if its center of mass is vertically above a point within the flat base. If the center of mass is vertically above a point on the earth that is not supporting the object, then the object will tip or roll.*

Center of mass: Throw a baseball bat, a hammer, or a ruler up in the air. The object will tumble but still move on a path close to the trajectory that a point mass would follow. In fact, one point that is fixed with respect to the object remains on that trajectory at all times, moving with acceleration **g**. That point is the center of mass of the object.

• Hang the object from a string and its center of mass will assume the same position that a point mass would: it rests vertically below the point of support. If it were off to one side, then the tension in the string and the weight, acting along different vertical lines, would produce a torque and the object would rotate. This fact can be used to find the location of the center of mass of any object. Attach the string to a few different points on the object's surface in turn, hanging it each time. In each trial, the center of mass lies on a line that is a downward extension of the line of the string. The intersection of the lines constructed in the different trials is the center of mass.

• The center of mass need not be inside the object. A boomerang, a carpenter's square, and a doughnut all have centers of mass near but outside the objects. Symmetric objects like the doughnut, a sphere, or a cube, have centers of mass at the points you would think of as their centers.

Stability:

• A soup can that is standing on a kitchen counter is stable. A vertical line through its center of mass intersects its circular base in the center. Set the can on a rough (so it will not slide) wood board and slowly raise one end of the board. As the angle from the horizontal up to the surface of the board becomes greater, the vertical line

35

through the can's center of mass moves away from the center of its circular base. While it is still within the circle, the board can exert an upward force on the downhill edge of the base great enough to make the net torque on the can zero. When the line has moved outside the circle, the upward force required at the downhill edge of the can's base to make the net torque on it zero exceeds the weight of the can. No such force is available. If there were such a large upward force, the can's weight would be too small to balance it; and the can would accelerate upward. Instead, the unbalanced torque causes the can to rotate downhill about an axis passing through the point of contact between the board and the downhill side of its base. The can tips over.

- Chairs and tables have four legs at the corners of rectangles as large as their supporting surfaces, so that placing even heavy objects anywhere on them does not make them unstable.
- Cars and trucks have wheels near the four corners for the same reason.
- Most mammals have four legs for effortless stability.
- We need feet that are long, front to back, strong muscles in the lower legs and feet, and a sense of balance—the constant, almost unconscious brain function that uses those muscles to keep the center of mass over the feet.
- A perfect cube of wood rests on the rough surface of a wooden ramp, with one edge perpendicular to the downhill direction. At what elevation angle of the ramp does the cube just begin to tip? (Answer: 45°)

OVERVIEW *Sample questions of the type that might appear on homework assignments and tests are presented with answers.*

Conservation of angular momentum:

- True or false: It is possible for the rate of rotation of an object to change even when there is not external torque acting on the object. True—Consider an ice skater who rotates rapidly with her arms pulled in, more slowly when she extends them. Her angular momentum is constant, so her rotation rate must change when she changes her momentum of inertia.

- True or false: It is possible for the angular momentum of an object to change even when there is no external torque acting on the object. False—The second law for rotational motion is that the rate of change of angular momentum is the torque, so zero torque means constant angular momentum.

Torque and balancing:

- Consider a meterstick supported by a string at the 50 cm mark. The mass of the meterstick is 0.01 kg. A 2 kg mass hangs from the 0 cm mark and a 1 kg mass hangs from the 75 cm mark. What is the tension in the string? The tension is equal to the total weight, $3.01 \text{ kg} \times 10 \text{ ms}^{-2}$ or 30 N.

- Where must you hang another mass if you want the system to balance, so that the meterstick of the previous question does not rotate? The torque tending to make the 0 cm mark drop is $2 \text{ kg} \times 10 \text{ ms}^{-2} \times 0.5 \text{ m}$ or 10 Nm. The torque tending to make the 100 cm mark drop is $1 \text{ kg} \times 10 \text{ ms}^{-2} \times 0.25 \text{ m}$ or 2.5 Nm. To balance, more mass is needed on the 100 cm side.

- At the 75 cm mark, what force must you apply to balance the system of the previous questions? You need to produce an additional torque of 7.5 Nm to make the torques on each side balance. At 0.25 m from the rotation axis, you must apply a force of 30 N down.

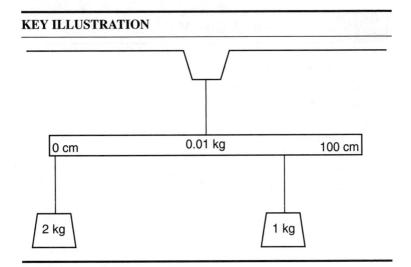

Theme 5 PLANETARY MOTION

This Theme presents the history of man's understanding of the motions of the moon, the planets, the moons orbiting other planets, and spacecraft. Newton's law of universal gravitation, combined with his laws of motion, explains and predicts planetary motions and motions of objects near the earth's surface. Before Newton, but even after Kepler had formulated his three laws that concisely state how planets move, it was not clear that one universal theory would be able to explain both terrestrial and heavenly motions.

OVERVIEW *Tycho Brahe measured where in the sky and at what time the "wandering stars" that we now know as other planets like earth appeared. He filled a library of notebooks with his data. Johannes Kepler, his successor, studied the data for years and learned that, in Copernicus's heliocentric (the sun is at rest at the center) frame of reference, planets obey simple rates of motion.*

The first law: Each planet moves in an elliptical orbit, with the sun at one of the two foci of the ellipse. An ellipse is a closed, curved line such that the sum of the two distances from any point on the line to the two fixed points, called *foci*, is a constant, the same for all points on the ellipse. That the orbits were not circles was a surprise to Kepler and his contemporaries.

- You can draw an ellipse using a length of string, two thumbtacks, and a pen or pencil. Make small loops in the ends of the string and fix them to the paper with the tacks. Leave some slack; do not stretch the string tightly. Place the pencil point against the string and move it as far from the tacks as you can. Press the point against the paper and move the pencil around the tacks, always keeping the string taut. You have drawn an ellipse with the tacks at the foci.

The second law: The line from the sun to a planet sweeps out or moves across equal areas in equal times. When a planet in a noncircular orbit is closer to the sun, it moves with higher speed than when it is farther away.

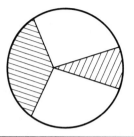

The third law: The square of the period of a planet's orbit around the sun is proportional to the cube of the average radius of the orbit:

$$T^2 \propto r^3$$

- The period is the time it takes the planet to go around the sun just once—one year for the earth.
- This law, like the first two, applies to the moons of Jupiter as well as to the sun's planets, but the constant of proportionality is different in the two cases.

Key 22 Newton's law of universal gravitation

OVERVIEW *Newton realized that the moon falls around the earth because of the same gravitational force that makes an apple fall to the ground when it is unsupported. He determined how the gravitational force must change with changing separation between two objects and how it must depend on their masses, if it holds planets in Keplerian orbits.*

The gravitational force: The magnitude of the gravitational force is

$$F_{grav.} = \frac{GM_1M_2}{r^2}$$

for masses M_1 and M_2 separated by distance r. G is a fundamental constant with the following value:

$$G = 6.67 \times 10^{-11} \text{ Nm}^2\text{kg}^{-2}$$

- The gravitational force acts along a line between the centers of the masses and is always attractive.
- Massive planets respond to the gravitational force just like any smaller mass—they accelerate according to Newton's second law.
- As you move up from earth's surface, the gravitational forces earth produces on objects and the local acceleration of gravity become weaker. People usually do not notice this because the height of the highest mountain is only about 10^{-3} times earth's radius. We cannot climb far enough to detect the change in $1/r^2$.
- But sending spacecraft into orbit, to the moon, or to other planets requires accurate use of Newton's $1/r^2$ law to find the forces acting on the spacecraft from earth, moon, sun, and some of the planets.

OVERVIEW *Mass, the measure of an object's inertia, depends on how much matter is in the object, but not on where it is located. Weight depends on where the object is located and even on the frame of reference in which an observer measures the object's weight.*

Mass: It is hard to avoid circular definitions in writing about mass and force. The value of the concepts is not reduced by this problem, however. Together, in Newton's theory, they explain so much and predict motions so accurately that we are willing to work with the theory and develop an intuition for the concepts from that experience. The modern, short definition is that mass is a quantitative measure of an object's resistance to any change in its state of rest or uniform motion. Newton described mass as being proportional to the amount of matter in an object. Both versions contain truth and help us think correctly about mass.

Weight: Near earth's surface the weight of an object is the gravitational force that the earth exerts on it.

- Near the surface of the moon or of one of the other planets, a small object's weight is the gravitational force on it produced by the nearby planet. The forces from other planets are negligible when one planet is near, because of the $1/r^2$ distance dependence and because the object and the nearby planet fall together toward any more distant planet.

- Your weight on the moon's surface would be about one sixth of your weight here, because the moon has much less mass than earth. On the moon, even if you had very sensitive instruments, you would not detect any tendency of the earth to pull you away from the moon. You would orbit the earth with the moon as earth's gravity makes you and the moon accelerate toward its center at exactly the same rate.

- In orbit in a space shuttle, you would feel weightless. The shuttle does not have enough mass to exert a noticeable gravitational force on you, and the earth makes you and the shuttle accelerate toward its center, in your circular orbits, at exactly the same rate. In a frame of reference that is itself accelerating, an object's weight seems to be different than in a frame of reference that is not accelerating with respect to the nearby planet. See also Keys 8 and 25.

Key 24 Gravitational fields

OVERVIEW *The preceding Key described a pattern of forces that one could measure by making individual force measurements at many points on the earth's surface. It is often convenient and useful to visualize such a pattern of forces, velocities, accelerations, displacements, or other quantities. A pattern of forces or accelerations is a force field or an acceleration field.*

Force and acceleration fields: At a single instant of time, the force produced by the moon acting on 1 kg of matter at a point on the earth's surface varies, in both magnitude and direction, from point to point. This variation can be shown in a single sketch. If a circle represents the earth, arrows at many points around its circumference can be drawn to represent the magnitude and direction of the force at each point. The sketch represents a force field—the force that would act on an object as a function of the object's location.

- The acceleration field has the same shape as the force field and the same size because the divisor used is 1 kg, but has units of m/s^2 rather than N.

The earth's gravitational field: The earth's gravitational influence, above and below its surface as well as at the surface, may be represented by a graph of the local acceleration of gravity g vs. r, the distance of the location from the center of the earth. At twice the earth's radius, 1.2×10^7 m from the center or 6×10^6 m above the surface, the acceleration described by Newton's universal law of gravitation is one fourth that at the surface. At three earth radii from the center, the acceleration is one ninth that at the surface; at four radii, one sixteenth; at five radii, one twenty fifth.

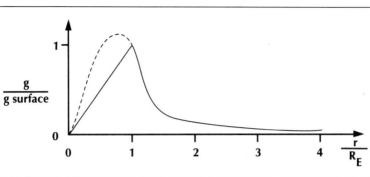

Inside the earth: How does the acceleration of gravity, g, vary as you move down beneath the surface of the earth? What force would the spherical shell of matter at radii greater than yours (at the bottom of a gold mine, say) exert on you down there? The two curves in the Key Graph that describe g below the earth's surface are mentioned in the last bullet in this Key and explained in Key 26. What gravitational acceleration does a uniform, thin, spherical shell of matter produce at a point p inside its inner surface? Imagine two narrow cones formed by lines that pass through the point (see Key Illustration following). There is a patch of the shell matter, a spherical cap, within each cone. Let the distances from the point to the caps be d_1 and d_2, with $d_1 + d_2$ the diameter of the shell. The area of the cap at d_1 is proportional to d_1^2, and so, therefore, is the mass in that cap. The gravitational force that cap 1 exerts on any object at the interior point is proportional to its mass and inversely proportional to the square of d_1 and is toward cap 1, away from cap 2. The same can be stated for the force produced by cap 2, except that its direction is opposite to that of the first force. The two forces are equal and opposite, because of the inverse-square nature of Newton's law. Move the point very close to one cap, and that cap becomes smaller while the opposite one grows. This effect on the mass in the caps just cancels the $1/d^2$ distance dependence of the force.

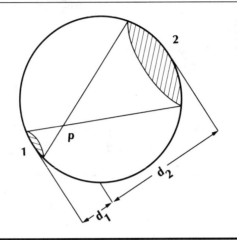

- Repeating this argument for other cones until all the mass in the shell has been considered produces the following result: the gravitational acceleration in a space enclosed by a uniform spherical shell of matter is zero. At the bottom of a gold mine, only the matter closer to the earth's center than you pulls on you.
- If the earth were a uniform sphere of mass, then g would begin to decrease below its surface (solid line in the graph). But g has been measured to increase as you go down (dashed line in the graph)— can you imagine why?

OVERVIEW *The moon's elliptical orbit is nearly circular. The acceleration needed to keep the moon on its circular path is exactly what Newton's law of universal gravitation predicts. No coincidence, this helped Newton realize that gravitational fields must have strengths that vary with distance* r *from the source mass as* $1/r^2$.

Acceleration in circular motion: An object moving along a circular trajectory at constant speed must accelerate perpendicular to the circle's circumference at every point.
- The directions of the velocity changes that keep the object on the trajectory require that the acceleration always point perpendicular to the circumference, toward the center.
- The magnitude of the acceleration toward the center of the circle of an object moving along at constant speed v is

$$a = \frac{v^2}{r}$$

where r is the radius of the circular trajectory.

The moon: The moon is about $d_{moon} = 3.8 \times 10^8$ m from the center of the earth, and the earth's radius is about $r_E = 6.4 \times 10^6$ m.
- Thus the moon is 60 times farther from the center of the earth than we are.
- The period T of the moon's orbital motion is about 28 days, or 2.4×10^6 s.
- The speed v of the moon in its orbit is the circumference of the orbit divided by the period, $2\pi d_{moon}/T = 10^3$ m/s.
- The acceleration of the moon toward the earth in its nearly circular motion is $a = v^2/d_{moon} = 2.6 \times 10^{-3}$ m/s^2, much less than the acceleration of gravity at the earth's surface, $g = 10$ m/s^2 (or, more precisely, 9.8 m/s^2).
- But Newton's law of universal gravitation states that the acceleration produced by the earth's gravity at the location of the moon is smaller than g:

$$a_{grav.} = g \left(\frac{r_e}{d_{moon}} \right)^2 = g/60^2 = 2.7 \times 10^{-3} \text{ m/s}^2$$

just what is needed to hold the moon in its orbit.

OVERVIEW *Sample questions of the type that might appear on homework assignments and tests, with answers.*

Kepler's laws:
- The radius of the orbit of Mars around the sun is 1.5 times the radius of the orbit of the earth around the sun. How many earth years does it take Mars to go around the sun just once? The period T, or duration of a planet's year, and the radius r of its orbit are related by $T^2 \propto R^3$, which means $T_M = T_E \times (r_M/r_E)^{3/2}$ or $T_M = 1.84$ years.
- An asteroid has an eccentric orbit. Its greatest distance from the sun is about 1.5 times its least distance from the sun. Its speed at its greatest distance is about what? The equal-areas law implies that the product of speed and distance from the sun is constant, so the speed at the greatest distance must be $\frac{1}{1.5} = \frac{2}{3}$ of the speed at the smallest distance.

Universal gravitation and gravitational fields:
- Oil is less dense than rock. Could you discover oil by its effect on g at the surface above it? Yes, the acceleration of gravity over oil would be slightly less than that over rock.
- Consider the question posed in Key 24. The answer is that $g(R)$ increases as R decreases below the earth's surface—as you descend a mine shaft for example—because the density (Key 29) of the earth is smaller at its surface than at its center. Descending, you reduce your distance from most of the earth's mass and eliminate the gravitational acceleration produced by only a small fraction of it. If the density of the earth were uniform, $g(R)$ would decrease as you descend.

Universal gravitation and gravitational acceleration:
- An astronaut sits in her super-shuttle 7×10^8 m above the center of Jupiter. The shuttle moves in a circular orbit with speed 1.4×10^4 m/s. What is its acceleration? This question can be answered using Newton's law of universal gravitation if Jupiter's mass can be found, or by using the acceleration of an object with speed v in a circular orbit of radius r, v^2/r. By the second method, $a = (1.4 \times 10^4 \text{ m/s})^2/(7 \times 10^8 \text{ m})$ or $a = 0.28$ ms^{-2}. Jupiter's mass is 1.9×10^{27} kg so the first method yields $a = 6.67 \times 10^{-11}$ Nm^{2}kg^{-2} $\times 1.9 \times 10^{27}$ kg/$(7 \times 10^8 \text{m})^2$ or 0.26 ms^{-2}.

STRUCTURE AND
PROPERTIES OF MATTER

This Theme presents the modern atomic model of matter and shows how the use of this model leads to an understanding of the common phases of matter: solid, liquid, and gas. Different solid and liquid materials in which atoms are packed closely together have different densities. A liquid, although its atoms or molecules slide by one another easily, resists compression forces. These properties produce buoyant forces on objects immersed in liquids, which were understood correctly by the Greek, Archimedes. Pressure and buoyancy occur in gases as in liquids. Pressure varies with the speed of a flowing gas; this fact is the basis of the science of aerodynamics.

Key 27 Atoms and molecules; elements, compounds, and mixtures

OVERVIEW *Pre-Socratic Greek philosophers speculated about the nature of matter. What would happen if one divided water into ever smaller portions? Would one obtain a smaller quantity of the same fluid at every step, forever, or would one find an invisible building block, Democritus's atom? The pre-Socratics hoped to find answers by pure thought, but could not. Scientists, formulating hypotheses, making predictions based on them, and then testing the predictions by experiments with matter, have learned that there are atoms.*

Atoms and molecules: The elements are hydrogen, helium, . . . , tellurium, . . .

- One atom of any element has unique properties characteristic of that element. Divide the atom (it is possible) and the parts no longer have any properties or behavior that would allow identification of the element they came from.
- An atom has a nearly spherical shape with a diameter of a few times 10^{-10} m—too small to see.
- Atoms may combine to form molecules. Simple examples of molecules are nitrogen (N_2), methane (CH_4), and water (H_2O). More complicated molecules containing many thousands of atoms each are found in living organisms, proteins for example.
- Molecules form and fragment in chemical reactions.

Elements, compounds, and mixtures:

- Bulk matter—solid, liquid, or gas—may contain only one element. Aluminum metal, argon gas, and the semiconductor silicon are examples.
- Compounds may be made of identical molecules, as is water (H_2O), or may contain atoms of different elements in specific proportions.
- Salt (NaCl) is a compound in which the atoms combine in the proportions implied by the chemical formula, but not by forming molecules that pack together.

- In much of bulk matter different elements are found together, but not in any particular proportion nor chemically bound together. These are mixtures of elements. Examples are the helium and neon gas in a laser cavity or 14 carat gold, an alloy of 58% gold and 42% silver and copper, proportions not specified by weight.

Key 28 The periodic table, atomic structure, and atomic masses

OVERVIEW *When ordered by mass from the lightest atom to the heaviest, the elements have chemical properties that vary from one atom to the next in a pattern that repeats itself. The atomic mass unit, AMU, has a convenient size for describing atomic and molecular masses.*

The periodic table: The repetition of chemical properties of groups of elements in a list sorted by atomic mass has inspired a table in which each row or period contains one copy of the repeating pattern.

- Atoms in a column of the table have similar chemical properties. For example, sodium, potassium, rubidium, and cesium all react violently with water and form compounds with any of fluorine, chlorine, bromine, and iodine.
- There are about 90 naturally occurring elements in the periodic table.

The structure of atoms: Most of the volume of an atom is occupied by electrons, which are relatively light particles with negative electrical charge.

- The changes in chemical behavior across a row of the periodic table are produced by the filling of different **shells** in the atom with electrons. The electronic charge density in an atom, as a function of distance from the center, has peaks corresponding to different shells separated by regions of lower density. See *Quantum Physics*, second ed., Robert Eisberg and Robert Resnick (New York: John Wiley & Sons, 1985), pp. 323–325.
- The mass of an atom is concentrated in its nucleus, which has a diameter 10^{-5} that of the atom.
- In liquids and solids, the separation of atoms corresponds to the size of the electron distribution, not the size of the nucleus. This fact suggests that forces between atoms are the electrical forces between their electrons.
- Neutrons and protons, both with mass about the same as the mass of a hydrogen atom, are in the nucleus. The proton has a positive electrical charge, and the neutron has no charge.

- Isotopes, which are atoms of the same element having different masses, differ only in the number of neutrons in their nuclei.
- The atomic number in the periodic table is smaller than the atomic mass. The atomic number is the number of protons in the atomic nucleus.
- The atomic mass, in AMU, is numerically very close to an average of the number of neutrons and protons of the isotopes of that atom found in nature. One AMU is exactly one twelfth of the mass of a carbon atom that has six protons, six neutrons, and six electrons. It is very close to the mass of a hydrogen atom.
- Ions are atoms with a number of electrons that differs from the atomic number. Ions are charged particles—negative if they carry an extra electron or positive if they are missing one or more electrons.

Solid structures and
their densities

OVERVIEW *In solids and liquids, atoms are packed
closely together so that their electron distributions just
begin to overlap. Interatomic spacings are typically 0.3 nm.
In solids the atoms may be packed in a regular structure
with identical, repeating units; in a nonrepeating but perfectly ordered structure; or they may be packed randomly.
Density is a property that depends on the atoms in a material but, for solids and liquids, not on the size or shape of a
particular sample.*

Crystals, glasses, and quasicrystals:
- A crystal is like a brick wall—a three-dimensional structure of
  identical units packed together that fills space with no gaps or
  overlaps. In a crystal, the packing units are atoms, molecules, or
  clusters of a small number of atoms in a particular spatial arrange-
  ment. Ordinary salt, sodium chloride—a crystal we see every
  day—displays its invisible atomic structure by the faceted surfaces of the crystals. All crystals tend to have flat surfaces parallel
  to planes of atoms in them.
- The atoms or molecules in glass are packed randomly rather than
  regularly. Window glass, randomly packed silicon dioxide, is a
  familiar example. Tape recorders and computer disk drives use
  glasses, made of metal alloys that have desirable magnetic properties, in their read-write heads.
- Quasicrystals are a new form of solid matter discovered at the end
  of 1984. In these metal alloys there are no single, repeating struc-
  tural units, but they can be as perfectly ordered as crystals. Interesting enough as an exotic new form of matter, they may also
  teach us about complex crystalline metal alloys that are important
  in aircraft and spacecraft components.

Density: Density is the ratio of the mass of a sample to its volume. The
density of water is 1 gcm^{-3}, because the mass unit, one gram, was initially defined as the mass of 1 cm^3 of pure water at a particular temperature. The SI density of water is 103 kgm^{-3}.

Liquids, pressure, and Pascal's principle

OVERVIEW *Near the earth's surface, a liquid flows to the bottom of its container. In the liquid closest to the bottom there are compression forces produced by the weight of the liquid above; this is the pressure in the liquid. Produce a change at one point in the pressure in a fluid that is at rest in a closed container, and the pressure at every other point changes by the same amount.*

Pressure: Consider the force F a liquid exerts on a small area A on the flat bottom of its container. It is perpendicular to the container wall because the liquid cannot resist shear (sliding) forces. It is small if the area is small, larger if the area is larger. The quantity that is constant, independent of the area, is the force per unit area. This is the pressure in the liquid:

$$P = \frac{F}{A}$$

- A liquid produces a force on an object immersed in it that is determined by the pressure where the object is.
- Pressure in a liquid with a free surface (a liquid-air interface) increases with increasing depth and is independent of the shape of the container.
- At any depth d the pressure P is just the weight of a uniform column of liquid of height d divided by the area of the base of the column:

$$P = \rho g d$$

where ρ is the density of the liquid and g is the acceleration of gravity. Strictly speaking, this is the amount by which the pressure at d exceeds the atmospheric pressure at the surface (Key 33).

- The SI unit of pressure is 1 Nm^{-2}, which equals 1 Pa or one pascal, named for Blaise Pascal, a French philosopher and natural scientist.
- The water pressure 1 m beneath the surface of a lake or pool exceeds that at the surface by 9.8 kPa.

Pascal's principle: Blaise Pascal first stated the simple rule that, in an enclosed fluid at rest, a change in pressure ΔP at any point is transmitted to all other points undiminished.

- The fluid pushes perpendicularly (no shear forces) against all walls of its container. Thus a fluid can transmit a force from one place to another, possibly with a direction change.
- The magnitude of the force on a moveable section of the wall, a piston, is proportional to the area of the piston.

Key 31 Buoyancy and Archimedes' principle

OVERVIEW *Because pressure in a fluid increases with depth, the fluid pushes up on the bottom of an immersed object harder than it pushes down on the top. The buoyant force on the object is the net force upon it produced by the fluid's pressure.*

Buoyancy: A cylinder immersed in a fluid with its axis vertical feels a force down due to pressure on its top, a larger force up due to pressure on its bottom, and forces on its sides that cancel. The net force is up.
- In this simple geometry, you can calculate the size of the net force in terms of the pressure and the geometry of the cylinder. It turns out to be proportional to the volume of the cylinder.
- For a complex shape the calculation becomes difficult.

Archimedes' principle: In approximately 250 B.C.E., Archimedes discovered the simple rule that describes the buoyant force on any immersed object, whatever its shape. The buoyant force has a magnitude equal to the weight of the fluid displaced by the object, and its direction is always up. This principle applies to objects that sink and that float.
- An object that sinks, completely immersed, experiences two forces, its weight W and the buoyant force F_b,

$$W = \rho_{object}\, g V$$
$$F_b = \rho_{liquid}\, g V$$

where V is the object's volume. You can measure both these forces with a spring balance. The object hanging in air pulls on the balance with W. The object hanging immersed pulls with $W - F_b$. The two measurements can be used with the formulae above to determine the mass and the volume of the object, whatever its shape, and so determine the density of the material. An object that sinks has a greater density than the density of the liquid.
- On a floating object, the forces W and F_b must be equal. A floating object has a density less than that of the liquid supporting it. Its total volume has a mass equal to the mass of a volume of water that is equal to the immersed part of its volume.
- Because an iceberg has a density about 0.9 times seawater density, 0.9 of the iceberg's volume is beneath the sea's surface.

OVERVIEW *A gas, like a liquid, is fluid; layers of material slide by each other freely. Unlike a liquid, a gas expands to completely fill its container, however large. The density of a gas is lower than that of a liquid or solid and is variable. A plasma is a gas in which some electrons have separated from their atoms leaving charged ions behind.*

Gases: In a gas, the forces that hold molecules together in condensed matter fail to keep the molecules close to each other. They fly about at speeds over 400 m/s (in air) colliding occasionally.
- If you take energy away from a gas, the speeds decrease and the forces between the molecules begin to make them stick together in low-speed collisions.
- Any gas will condense to a liquid or to a solid if its energy is reduced.

Boyle's law: In the seventeenth century Robert Boyle discovered that if temperature and total number of molecules in a gas are held constant, the product of pressure times volume of the gas is constant:

$$P_1V_1 = P_2V_2$$

Stated differently, for a gas at fixed temperature, pressure is proportional to density.
- To increase the pressure in your automobile tire, you must put more air into the fixed volume, which increases the density of the air inside the tire.

Ideal gas law: The dependence of Boyle's pressure-volume product on the temperature T and the number of molecules in the container N is simple, for gases well above the temperatures at which they condense to liquids:

$$PV = NkT$$

in which k is a constant with dimension energy/temperature.
- The pressure in your automobile tire increases during the day, as increasing air temperature and frictional forces during driving increase the temperature of the tire.

Plasmas: In a plasma there are negative electrons separated from atoms, positive ions (atoms that lost electrons), and negative ions that have captured extra electrons.

- A plasma conducts electricity.
- Fluorescent lamps, many street lights, and neon signs contain plasma through which an electric current passes. When electrons and ions recombine, the atoms they form often emit light.

Key 33 The earth's atmosphere and its pressure

OVERVIEW *Our atmosphere, roughly 10 km high, is mostly nitrogen (N_2) gas. Its weight produces great pressure at sea level.*

The atmosphere: Besides N_2, the atmosphere contains some oxygen gas (O_2) and carbon dioxide gas (CO_2), some water vapor, a bit of argon (Ar), and traces of other gases. Gravity keeps it here and produces the density.

- The density of the atmosphere varies with temperature and altitude and at sea level is about 1.2 kgm^{-3} or 1.2×10^{-3} gcm^{-3}, roughly 10^{-3} the density of water. Half of the molecules in the atmosphere are below a height of 6 km; 90 percent are below 20 km; and 99 percent are below 30 km.

Pressure: Just as in liquid, the weight of the air above us produces an atmospheric pressure all around us. It is 10^5 Pa (Key 30). We are not crushed and do not even notice, because the pressure inside us is the same as the atmospheric pressure outside. But if the air from inside a closed container is removed, the effect of pressure on the outside of the container becomes more obvious.

- An evacuated gallon gasoline can will collapse under the unbalanced outside pressure.
- In 1654, Otto von Gueicke of Magdeburg placed two hemispheres together with a sealing gasket and used a vacuum pump he had made to remove the air from the sphere, which was less than 1 m in diameter. Two teams of horses could not pull the hemispheres apart until he let the air back in.
- A **barometer** measures atmospheric pressure. It has uses in weather prediction and in determining altitude from the pressure. In one version, the pressure of the atmosphere supports a column of liquid above which there is a vacuum. The height of the fluid column is proportional to the pressure at its base. The atmosphere can support a column of mercury (Hg) 760 mm high, or a 10 m water column. When you drink through a straw, atmospheric pressure pushes the drink up into the partial vacuum you create. Water wells more than 10 m deep cannot be pumped from the top. Deep wells have pumps placed at the bottom to push up the water column.

Key 34 Buoyancy of air; Bernoulli's principle

OVERVIEW *In a stationary fluid near the earth's surface, atmospheric pressure decreases with increasing altitude (Keys 30 and 33). This produces a buoyant force on objects immersed in the atmosphere. Pressure also varies with the speed of a moving fluid.*

Archimedes' principle: Archimedes' simple rule works for buoyant forces produced by gases as well as for those produced by liquids. The force on an object immersed in the earth's atmosphere is equal to the weight of the air that the object displaces.

- A sphere 30 cm in diameter near sea level is pushed upward by the air with a force of 0.17 N. If the sphere is a helium balloon with a mass less than 0.017 kg, or 17 g, then the buoyant force is greater than the weight, and the balloon will rise.
- Is the Goodyear blimp less massive, more massive, or equal in mass to the mass of the air it displaces? Because the buoyant and gravitational forces on the blimp are equal and opposite, the blimp must be equal in mass to the mass of the displaced air.

Pressure and speed: In the eighteenth century, Daniel Bernoulli, a Swiss scientist, noticed that fluids flow faster when forced through narrow constrictions. A river with alternating broad pools and narrow disconnecting channels is an excellent example if you can find one to observe.

- Bernoulli reasoned that when kinetic energy increases, another form of energy must decrease. High pressure in a fluid stores elastic energy, just like the compression of a spring. Pressure must decrease as speed increases in a moving fluid. This is Bernoulli's principle.
- There are many examples of lower pressure in flowing air or water. Some are useful and others are harmful to us. An airplane wing is shaped so that air passing over it must travel further than air passing under it. The air above the wing moves faster and therefore exerts a lower pressure on the wing than the air below. This produces the lift force that keeps planes up. High winds in a storm passing over a house produce a pressure on the windows and roof that is lower than atmospheric pressure in still air. The air inside the house may break windows or push the roof off.

OVERVIEW *Sample questions of the type that might appear on homework assignments and tests are presented with answers.*

Atomic structure:
- The mass of matter comes mostly from what? What determines the sizes of atoms? The mass comes from the neutrons and protons in the nucleus; the size is determined by the extent of the spatial distribution of electrons around the nucleus.

Density:
- If the mass of an object were to double while its volume remained the same, how would its density change? Density, mass/volume, would double.
- A metal block has a density ρ of 5×10^3 kgm^{-3} and a mass m of 1.5×10^4 kg. What is its volume V? $V = m/\rho$ or 1.5×10^4 kg/ 5×10^3 kgm^{-3} or 3 m^3.

Buoyancy:
- Against what part of a completely submerged object is water pressure greatest? Water pressure increases with depth and so is greatest on the bottom of a submerged object.
- A 1 kg block of lead and a 1 kg block of aluminum are completely submerged in water. On which block is the buoyant force greatest? The less dense aluminum must have a greater volume than the same mass of more dense lead, so the aluminum displaces more water. The buoyant force is greater on the aluminum.
- A car with closed windows turns left. Which way would a helium-filled balloon in the car move? Dense air resists the acceleration to the left more effectively than the less dense balloon; the balloon moves to the left.

Aerodynamics:
- Which is the most important physical principle for the flight of an airplane—Archimedes' principle, Bernoulli's principle, or Pascal's principle? It is Bernoulli's principle that explains how wings provide lift and how propellers—small rotating wings—provide thrust.

TEMPERATURE, HEAT,
AND THERMODYNAMICS

This Theme begins with two results from mechanics. First, mechanical energy is lost in motion with frictional forces. Heat has been considered a form of energy to retain the notion of conservation of energy. Second, in the atomic model of matter, the atoms may vibrate (condensed matter) or fly about (gas or plasma) randomly and thus have kinetic energy. Heat flow and the definition and measurement of temperature are discussed. Melting and boiling occur at particular temperatures. We shall describe how the changes occur. The first two laws of thermodynamics tell how heat flows and how it can be changed into other forms of energy. The new concept of entropy is introduced.

OVERVIEW *The kinetic energy of the random motion of molecules in matter makes matter seem hot; increasing this thermal energy increases an object's temperature. When thermal energy leaves an object and flows to another object, we say that heat flows. Temperature measures the tendency for heat flow.*

Temperature: Temperature is a quantitative measure of how hot or cold an object feels to us.

- Anders Celsius, an eighteenth-century Swedish astronomer, assigned 0 to the temperature at which ice melts and 100 to the temperature at which water boils. Later users of his scale have exchanged the values. G. D. Fahrenheit, a seventeenth-century German scientist, assigned 32 and 212 to the same temperatures. The nineteenth-century British physicist Lord Kelvin based his temperatures on the observations of Jacques Charles that suggested that there was a temperature at which the kinetic energy of the molecules in a gas became zero. Kelvin assigned 0 to this and, following Celsius, divided the range from freezing to boiling of water into 100 parts.

- Celsius, Fahrenheit, and Kelvin scales are in common use today. The units of temperature are 1°C or 1 K for the Celsius or Kelvin scaled. 0 K is –273°C, the absolute zero at which gas molecules should stop moving. A change of 1°C is the same as a change of 1 K, or the same as a change of $\dfrac{180}{100} = \dfrac{9}{5}$ Fahrenheit degrees. 0°C is 32 Fahrenheit degrees.

- One liter or 100 cm^3 of boiling water has 1000 times as much thermal energy as 1 cm^3 of boiling water, but they both have the same temperature.

- One liter of water just above its freezing temperature has **more** thermal energy than 1 cm^3 of boiling water.

- A temperature difference between two objects, whatever their size, means that heat would flow from the hotter object to the colder object if they were in thermal contact. If the freezing and boiling water just described were mixed, heat would flow from the 1 cm^3 at 100°C to the 1 L at 0°C, even though the total thermal energy initially in the 1 L was greater.

OVERVIEW *Heat units are the amounts of heat required to produce standard increases in the temperatures of standard substances. But it takes more heat to raise the temperature of 1 g of water 1°C than it takes to raise the temperature of 1 g of steel 1°C. Specific heats describe these differences between materials. In the nineteenth century, Count Rumford observed that, like heat flow, work against friction forces increases thermal energy.*

Units of heat: One calorie (1 cal) is the heat required to increase the temperature of 1 g of water by 1°C. One Calorie (1 Cal) is 1 kcal, distinguished only by the capital letter, and is the unit commonly used to measure the energy content of foods.
- One calorie will raise the temperature of 1 g of most substances other than water by more than 1°C.
- The specific heat of a material is the number of calories required to raise the temperature of 1 g by 1°C.
- Water has a greater specific heat, or capacity for holding heat, than other substances.
- With a given temperature change, 1 g of water takes up more heat than 1 g of most other things.
- Both steam and ice have specific heats of around ½; steel has a specific heat of about ⅛.

The energy in one calorie: To determine the mechanical equivalent of heat, one must measure the work done in a situation in which all work is converted to thermal energy in a measurable and well-insulated amount of material. James Joule determined that 1 cal = 4.19 J in such an experiment.
- The energy that raises the temperature of water 1°C could instead elevate it 428 m above its starting position, if delivered as mechanical work rather than heat.

Key 38 Thermal expansion

OVERVIEW *With a few interesting exceptions, materials tend to expand as their temperatures rise. Structures must be designed with this in mind to avoid thermally induced breaking. Some useful devices are based on this phenomenon.*

Jiggling molecules: In condensed matter, the increase in kinetic energy of molecules with rising temperature means that the molecules collide harder and push on each other harder. Like a good basketball player with elbows out and moving, these faster-moving, harder-hitting molecules take up more room.
- Both liquids and solids expand as their temperatures rise.

Designing for expansion: Concrete roads and sidewalks are rigid and inflexible, yet they expand when they are heated. That is why they are made with empty gaps in sidewalks or tar-filled gaps in roads. As the concrete expands, the gaps shrink, but never so much that the blocks of concrete press against each other. Long bridges have the steel-and-concrete roadway supported on rollers on at least one end. There are also expansion joints in the surface with interlocking steel surfaces connected to adjacent road sections, so vehicles are supported even on cold nights when the joints open (See Key Illustration).

Applications: Glass thermometers indicate temperature by the expansion of a liquid contained in a bulb at the bottom of a very thin glass column. The contained fluid, either mercury or colored alcohol, must move up the column as it expands. (What would happen when the temperature rises if the glass expanded more rapidly than the liquid?)
- Thermostats to control home heating and cooling systems are often made from bimetallic strips—thin strips of two different metals bonded together. The metals are selected so one expands more than the other with increasing temperature. Different expansion on the two sides produces curvature. The effect can be amplified by winding a long strip into a coil. The free end of the coil, moving with temperature changes, can move an indicator dial or change the state of an electrical switch.

Water, an exception: Liquid water expands as its temperature rises above 4°C (about 39° Fahrenheit), but water also expands slightly as its temperature decreases from 4°C to 0°C. The fact that ice floats helps us understand this. With a density 0.92 that of water (Key 29), ice has more space between the molecules than water does. This is a consequence of the way water molecules orient as they bond to each other. This orientation forces some openings in the ice crystals.

- In almost-freezing water, intermolecular forces produce transient clusters with the structure of ice. These clusters melt almost as soon as they form, but they persist long enough to produce the 2×10^{-4} fractional volume expansion from 4°C to 0°C (the expansion on freezing is 0.08).

- Many bottom-dwelling freshwater plants and animals survive winters in ponds because of the exceptional behavior of water. Water at 4°C is the densest, and sinks to the bottom as a pond cools. The near-0°C water rises, and so freezing begins at the surface. Bottom water is the last to freeze.

KEY ILLUSTRATION

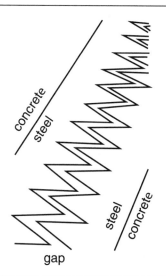

Key 39 Newton's law of cooling; heat transfer

OVERVIEW *Newton discovered and stated how the rate of heat flow from a hot object in thermal contact with a colder object depends on the temperature difference. Heat can flow between bodies in thermal contact, but what constitutes thermal contact? This Key looks at the ways heat flows.*

Newton's law of cooling: The rate at which heat flows from a hot body to a colder body is proportional to the difference in temperature between the two bodies. This is Newton's law of cooling.

- A hot cup of coffee will lose heat rapidly at first, more slowly when it has cooled some. Its temperature drops more rapidly when it is hot than when it has cooled.

Contact: When two different materials are in contact, jiggling surface molecules in one can transfer kinetic energy to molecules in the other by direct, molecule-against-molecule collisions. That is why your feet feel cold as you stand on the bathroom floor after a shower. Fast-moving foot molecules lose energy in collisions with slower-moving floor molecules. But why does the foot on the tile floor feel so much colder than the foot on the rug? Why does the tile carry energy away from the foot more rapidly? Specific heat is part of the answer but not the whole answer. After all, there is a tile beneath the rug just like the one directly beneath the foot not on the rug.

Conduction: Different materials conduct heat with different efficiencies.

- You can light a wooden match and hold it comfortably until the wood burns away and the flame approaches your fingers. But hold a nail in your other hand and place its far end in the flame; you will have to drop the nail long before you have to drop the match. Metals are good conductors of heat. A little random kinetic energy added at one end spreads quickly throughout a nail. This is because electrons can move freely through metals. In their collisions with each other and with atoms, they are very effective in transporting energy.
- Wood, cloth, and plastic have no such itinerant electrons and therefore insulate or restrict the flow of heat. Handles on saucepans and barbecue grills are made of plastic or wood to keep

their heat from our hands. Cloth rugs slow the flow of our heat to the cold floor.

Convection: Liquids and gases expand as their temperatures increase. Near the earth's surface this produces natural convection currents. The warm, lower density material rises and colder, denser material sinks.

- Forced convection can be driven by fans and pumps if gravity and density changes do not do the job desired.
- A moving fluid carries its thermal energy with it. If hot fluid moves one way and cold the other, then energy is flowing in the direction the hot fluid moves.
- Convection in air and in water transfers heat much more efficiently than conduction would in stationary fluids.
- Insulation for homes, for people in cold air, and for people in cold water has a number of common features. First, conduction by the insulating material can be reduced if the amount of material is reduced. Insulators are designed to use a minimum of the insulating material, reducing conduction while still preventing direct contact between cold air or water and the person or home. Air or water which fills in the gaps or pores in the insulator, have relatively low conductivities. Finally, the gaps or pores are made very small, which inhibits the formation of convection currents.

Radiation:

- Hot matter, even isolated in the middle of a vacuum, loses energy to the cool walls of the vacuum chamber. The heat is transferred by electromagnetic radiation (Key 78).
- Two otherwise identical objects at the same temperature, one black and one shiny, absorb and emit radiation differently. The black one absorbs most of the radiation that falls on it; the shiny one reflects most of the radiation away. The black one must emit more radiation than the shiny one. If it did not, it would warm up while the shiny one would cool. White snow on the ground reflects sunlight away during the day and keeps the ground from warming up. At night the snow radiates more slowly than would the dark ground and so slows the rate of its cooling.

OVERVIEW *A heated liquid turns to gas; the gas, confined and cooled, turns back to liquid.*

Evaporation: Molecules in a liquid move at high speeds, run into each other frequently, and move on again at high speeds. Some molecules move faster, others slower. The average speed depends on the temperature. At any temperature, some molecules move faster than the average. Near the surface of the liquid, some molecules move toward the surface and others move away from it. Molecules near the surface, moving toward the surface, and moving fast may escape from the liquid. If the attractive forces from other molecules in the liquid that are strong enough to hold slow molecules back are not strong enough to stop the fastest, some fast molecules leave the liquid. Molecules with the greatest kinetic energy leave the liquid, and those with lower kinetic energy stay.

- Evaporation is a cooling process. As evaporation proceeds, the average kinetic energy per molecule remaining in the liquid decreases. The liquid cools; its temperature decreases. If you do not believe this, just lie on your back and pour a bit of rubbing alcohol on your bare tummy. What happens as it evaporates?

- We sweat to cool our bodies in hot weather. We exude water onto our skin. As the water evaporates, it carries away heat and cools us. On a humid day when there is so much water in the air that sweat is slow to evaporate, we feel hot and wet.

Condensation: Gas molecules colliding with a cool surface may lose energy in their collisions. Sometimes, they may lose so much energy that they cannot move away again—they stick. As more and more molecules stick to each other, a liquid layer grows on the surface. This is condensation. It can proceed only if there is some place where the greater kinetic energy of gas molecules can go as they enter the lower kinetic energy liquid phase.

- Condensation occurs readily on a cold surface near a dense gas but not on a hot surface in the same gas. That is why car windows are such a mess on a cold morning and why heating them with hot air from a defroster helps.

- Air always has some water vapor in it. As air cools, water molecules begin to stick together. This process forms clouds high above the ground and fog near the ground: small, suspended drops of water.

Boiling: Heat a liquid, like water, and more and more of its molecules gain enough energy to escape from the surface and enter the gas phase. With enough energy, escaping molecules can push hard enough against the liquid and the atmospheric pressure above it to push the liquid aside and make a bubble. This is boiling. If atmospheric pressure above the liquid increases, more energy is needed to make a bubble; the boiling temperature increases.

- If you climb a mountain and reach a height where atmospheric pressure is lower than at sea level, the temperature required for boiling decreases.
- Pressure cookers cook food faster because they maintain higher pressures, which increase the boiling temperature.
- Mountain campers must cook food longer because their boiling water is not as hot as it would be at sea level.
- Like evaporation, boiling cools the liquid left behind. We do not usually notice this because the water we boil is over a heater that adds energy as fast as the evaporation removes it.

Key 41 Melting, freezing, and sublimation

OVERVIEW *In a liquid, molecules stay close together but change places and slide by each other. In a solid, molecules stick more tightly and no longer change places. Attractive forces between molecules freeze them into fixed positions when their average kinetic energy is low.*

Melting and freezing: Slowly vibrating molecules can move back and forth without ever moving away from the nearby molecules that hold them tightly. In a solid at nonzero temperature, nothing holds still, but most molecules stay close to their average positions. Add more kinetic energy (heat), and molecules begin to vibrate so vigorously that they slide by each other easily. This is melting. Freezing is the inverse process to melting. Remove heat from a liquid, and its molecules begin to stick together (i.e., stop sliding by each other). A solid forms.

Sublimation: Under some conditions, molecules go from the solid state to the gas state, never becoming liquid. The most familiar example of sublimation occurs with snow on dry days with temperatures below 0°C. The snow disappears without any liquid appearing. Molecules gain energy and escape from the solid, not to a state stuck together and free to slide by each other (liquid), but directly to a state in which they do not stick together (gas). Molecules with the highest energies escape. The snow left behind becomes cooler, unless the sun heats it as fast as sublimation cools it.

Key 42 Energy in changes of state

OVERVIEW *Melting and boiling occur suddenly, rather than continuously. Boiling occurs only when the energy of evaporating molecules is strong enough to oppose atmospheric pressure and form a bubble. Amazingly, this happens at one temperature, as does melting.*

Boiling: Heat a liquid until its molecules have almost enough energy to break away from each other, and then what? The temperature has increased with the added heat. But adding still more heat does not increase the temperature further. Instead, it gives molecules the energy they need to leave the liquid. More and more added heat drives off more and more liquid molecules, without changing the temperature of the liquid that stays behind. Only when all the liquid molecules have separated from each other does the temperature of the gas begin to increase with further additions of heat.

Heat of vaporization of water: To change 1 g of water already heated to 100°C from a liquid to water vapor at 100°C, you must add 540 cal of energy. To vaporize 20 g, you must add 20×540 cal or 1.08×10^4 cal. This is much more heat per gram than the 100 cal required to heat 1 g of water from 0°C to 100°C. To condense 1 g of steam at 100°C to liquid water at 100°C, you must carry away 540 cal of heat.

Heat of fusion of water: To freeze 1 g of liquid water at 0°C to ice at 0°C, you must remove 80 cal of heat. To melt ice at 0°C to liquid water at 0°C, you must add 80 cal of heat. This is the heat of fusion of water. Melting and freezing, like boiling and condensation, occur at one fixed temperature. As you add heat slowly to ice, for example, it warms up to 0°C. Then, with more heat, it remains at 0°C as some of it melts to water. The material, ice and water, remains at 0°C until all the ice has transformed to water; only then does the temperature of the water rise again with further addition of heat. This also happens with freezing as heat is slowly removed.

Key 43 The first and second laws of thermodynamics; entropy

OVERVIEW *Thermal energy is the energy of the random motion of molecules in a material. Heat flow is the transfer of thermal energy from one object to another. In this Key, conservation of total energy is carefully extended to cover conversions between thermal energy and other forms of energy. There are many different ways of stating the second law. It may not be obvious to students that they are all equivalent, but they are.*

Heat is flowing thermal energy: Heat added to a material transforms to some form of energy—either thermal (internal) energy contained by the material or some other nonthermal energy.

- Work can produce thermal energy, and thermal energy can produce work.
- The heat Q added to a system (like a steam engine) or a material (like ice in a sealed metal container) divides in some way into an increase in the internal energy U and some work W done by the system. Conservation of energy requires that

$$Q = \Delta U + W$$

This is the first law of thermodynamics.

Heat flow: One statement of the second law is: heat always flows from a hotter object to a colder object, never the other way. From our experience of putting hot and cold things together, such as ice cubes and water, this seems obviously true. But could there be some way? Sadi Carnot, a nineteenth-century French scientist, said, "No."

Heat engines: After James Watt in England made a steam engine efficient enough to be worth using (instead of a horse or a man) in pumping water out of mines, Carnot undertook a study of the efficiency of heat engines: how much work could a heat engine produce, per unit of heat energy put into it? He proved that there was a maximum possible efficiency such that a more efficient engine would make possible a violation of the form of the second law stated in the previous paragraph; a too-efficient heat engine could be made part of a complicated connection between a cold place and a hot place that would make heat flow from cold to hot without any external energy input. If you do not expect this to occur, then any engine that takes in heat from a hot place must deliver some of it to a colder place; it cannot convert all of its heat input to mechanical work.

- Carnot's efficiency is defined as

$$\epsilon = \frac{\text{work out}}{\text{heat in}}$$

According to the second law, the greatest possible efficiency for a heat engine is

$$\epsilon_{\text{ideal}} = \frac{T_{\text{hot}} - T_{\text{cold}}}{T_{\text{hot}}}$$

where T_{hot} is the temperature of the place from which heat enters the engine, and T_{cold} is the temperature of the place to which heat is ejected. These temperatures must be absolute, measured on Lord Kelvin's scale or another like it. For example, an automobile engine takes heat from gasoline burning at a very high temperature and rejects heat out its exhaust system at a lower temperature not too much greater than room temperature. The high temperature from which heat is delivered to the engine is something like 2200°C or 2500 K. Exhaust gas is hot, but not that hot. If the exhaust leaves at 200°C or about 500 K, then the maximum possible efficiency, the Carnot efficiency, is

$$\epsilon = \frac{2500 \text{ K} - 500 \text{ K}}{2500 \text{ K}}$$

or $\epsilon \approx 0.8$. At most, 80 percent of the chemical energy in the gasoline can be transformed to work done against the wind, internal friction, and road friction. At least 20 percent must be blown out the exhaust as wasted heat.

- Some chemical energy is useful (work) and the rest is wasted (heat at low temperature) in a heat engine. Jumping to the microscopic model of matter, this means that some, but not all, of the random kinetic energy of molecules in matter that is heat can be transformed into nonrandom kinetic energy of human-scale parts of engines. You cannot take a group of molecules whizzing every which way and persuade them to move left, all at the same time. Randomness, lack of organization, cannot be completely undone.

Order and disorder: Isolated systems tend toward a greater degree of disorder. This is the next-to-last form of the second law considered here.

- A small number of molecules with violent random motions has disorder. A large number of molecules with less-rapid, random motions may have more or less disorder. Sadi Carnot could have done more if he had had a quantitative measure of disorder. J. Willard Gibbs had one and did much more; he invented the field of statistical mechanics.

- The quantitative measure of disorder, entropy, is analogous to mass, which is the quantitative measure of inertia. The final statement of the second law here is that isolated systems always tend toward a state of greater disorder. The entropy of an isolated system remains constant (at best) or increases.

- As an example of the increase of entropy, consider a container of gas with a wall across the middle. Let all the gas be on the left side of the wall, with a vacuum on the right. With this information, we know there is some small degree of order in the system. Remove the wall slowly and the information is lost; the order is lost. You cannot put the wall back and recover the initial state of the system. Its entropy has increased. A desk, although never really isolated from outside influences, behaves in some ways like a thermodynamic system. Has your desk ever spontaneously become neater?

OVERVIEW *Sample questions of the type that might appear on homework assignments and tests are presented with answers.*

Molecular motion and temperature:
- In a mixture of hydrogen, nitrogen, and oxygen gases at room temperature, which molecules have the greatest average speed? Each type of molecule has the same average kinetic energy, so the least massive, hydrogen, must have the greatest average speed.

Heat transfer:
- Do objects that radiate relatively well absorb radiation relatively well, reflect radiation relatively well, or both? They absorb well, reflect relatively little.

Boiling:
- Does the increased air pressure on the surface of hot water tend to prevent boiling, promote boiling, or neither? It tends to prevent boiling by raising the energy required to form a bubble within the fluid and so raising the boiling temperature.

Thermodynamics:
- Two identical blocks of iron, one at 10°C and the other at 20°C, are put in contact. Assume they are well insulated from the rest of the world. Suppose the cooler block goes to 5°C and the warmer block goes to 25°C. Would this violate the first law of thermodynamics, the second law of thermodynamics, or neither? This would violate the second law, with a heat flow from a cold object to a hot object, but not the first.
- If you run a refrigerator with its door opened in a closed room, will the room temperature decrease, increase, or remain the same? A perfect Carnot heat engine would pump out more heat to the high-temperature heat reservoir, the room air, than it absorbs from the low-temperature reservoir; it would make the room hotter. A less perfect real refrigerator would heat the room faster than Carnot's ideal engine.
- What is the maximum possible efficiency ε of an engine operating between 227°C and 27°C? It is $(T_{hot} - T_{cold}) / T_{hot}$, with temperatures on an absolute scale. Using Kelvin's scale, $\varepsilon = (500 \text{ K} - 300 \text{ K}) / 500 \text{ K}$, or $\varepsilon = 0.4$.

Theme 8 OSCILLATIONS, WAVES, AND SOUND

This Theme describes periodic motion, which is motion that occurs and repeats itself, again and again. A wave is a disturbance that travels through a medium. A periodic wave is a motion that repeats itself in space as well as time. The mechanical picture of a sound wave in a gas is developed, and the relationship between perceived qualities of sound and physical properties of waves is given.

INDIVIDUAL KEYS IN THIS THEME

Key 45 Oscillation of a pendulum

OVERVIEW *A pendulum swings back and forth, repeating the same motion over and over. This periodic motion is called oscillation. Galileo discovered the simple relationship between the period (time for one swing) and the physical properties of the pendulum.*

The simple pendulum:
- The length *l* of a pendulum is the distance from the point of support to the center of mass.
- Galileo, who observed that large and small masses fall with the same acceleration, also noticed that the period of a simple pendulum, which is the time for one cycle of its motion, is independent of the mass.
- For small **amplitudes**, which are swings through arcs of only a few degrees, the period is independent of the amplitude.
- The period depends on the length *l* of the pendulum and on the acceleration of gravity *g*. The relationship is given as

$$T = 2\pi \sqrt{\frac{l}{g}}$$

- A 1 m pendulum has a period of about 2 s when it swings through an arc of only a few degrees.
- A grandfather clock uses a 2 s pendulum and gears arranged so the second hand advances one s click for each half-swing of the pendulum.

Acceleration of gravity: Newton's law of universal gravitation (Key 22) shows that a large region of lower-than-average-density rock beneath an observer may make *g* smaller at the observer's location than at other locations. Precise measurements of *g*, using instruments based on pendulums, help locate important, low-density minerals like oil.

Simple harmonic motion: The motion of a pendulum swinging through an arc of only a few degrees is important in physical science. It is found in many different systems and so has been given the name *simple harmonic motion.*

- A graph of the position of the pendulum bob vs. time is a sine or cosine function of time, $\theta = \theta_0 \sin(2\pi f t)$ for example, in which θ is the angle of the pendulum away from the vertical and f is its frequency. These harmonic functions appear in the descriptions of periodic waves.

KEY ILLUSTRATION

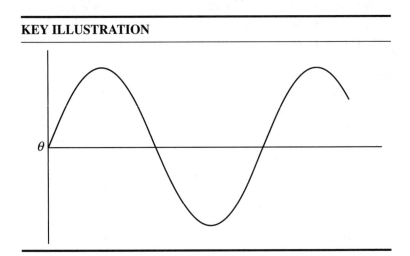

Key 46 Description of a wave: motion and speed

OVERVIEW *A wave is a disturbance in a medium that moves through the medium at a constant speed. Amplitude is the size of the disturbance. In this Key, we concentrate on periodic waves and the quantitative parameters used to describe them.*

Velocity and amplitude: Water waves are usually periodic waves that move with a well-defined velocity. The amplitude is half the vertical distance from the bottom of a trough to the top of a crest.

Period and frequency: The period T of a wave, like the period of a pendulum, is the time between repetitions of the periodic motion. In a water wave, the period is the time between crests at a selected location. The frequency is the number of cycles of the motion that occur in a unit time. The frequency f is measured in cycles per second or Hertz (Hz), named after the German Heinrich Hertz who first detected radio waves in the late nineteenth century. The relationship between period and frequency is as follows:

$$f = \frac{1}{T}$$

- Water waves have frequencies around 1 Hz.
- Radio waves have frequencies around 1 MHz (AM) or 100 MHz (FM).
- Light waves have frequencies around 0.6×10^{15} Hz.
- Sound waves have frequencies from a few tens to a few thousands of Hz.

Wavelength: Wavelength is the distance at one instant of time between two adjacent wave crests. The relationship among wavelength, propagation speed v, and frequency f is as follows:

$$v = f\lambda$$

- Sound travels in air at about 340 m/s.
- Light travels with the speed 3×10^8 m/s.
- The wavelength of sound in air is typically 1 m.
- Radio waves have wavelengths of 300 m (AM) or 3 m (FM).
- Light has wavelengths around 5×10^{-7} m.

Key 47 Longitudinal and transverse waves; mixed waves

OVERVIEW *When a wave passes through a medium, which way does the material in the medium move? The answers are different for different kinds of waves considered here.*

Longitudinal polarization: Stretch a Slinky® spring out to a length of about 2 m on the top of an empty table (you may need a friend to hold the other end). Rapidly move your end of the spring toward your friend and then back toward you. This should produce a solitary wave traveling away from you. As it passes a point, the spring coils move in a direction parallel to the wave's velocity. This is a longitudinally polarized wave.

- Sound waves in liquids and gases are longitudinal; the resistance of liquids and gases to compression forces allows the waves to propagate.

Transverse waves: Facing your friend, who is holding the other end of a spring on a table, move your end of the spring rapidly left and then right. This should produce a solitary wave traveling away from you with transverse polarization. As the wave passes a spring coil, the coil moves in a direction perpendicular to the wave's velocity. In three dimensions, there are two perpendicular transverse polarizations possible. For your spring, the other one is in the vertical direction, but it is dampened by the table supporting the medium.

- Electromagnetic radiation is always transversely polarized.

Mixed waves: Water waves are neither longitudinal nor transverse, but a mixture of the two. As a wave passes you, the water moves in an elliptical path in a vertical plane. It surges forward and up as a crest approaches, down after it passes, and backward in the trough.

- When a wave passes through a medium like water or air, the wave can travel a long distance. None of the matter in the medium moves very far, however. Every molecule is close to its starting position after the wave has passed.

Key 48 Interference and standing waves

OVERVIEW *If two simple harmonic waves pass through a single point, the displacement of the medium (at that point) is the sum of the displacements each wave would produce acting alone. Although there are interesting exceptions to this rule, there are many situations in which it holds true. Superposition of two single waves gives the description of the motion of the medium when both waves are present.*

Interference: When two waves pass a single point they are said to **interfere**. Their amplitudes may add to give a larger amplitude at the point—**constructive** interference; or their amplitudes may cancel (subtract) to give a small amplitude—**destructive** interference.

Standing waves:

- Tie a rope to a doorknob or drawer handle and move back, taking up any slack but not stretching it too tightly. Move your end of the rope up and down continuously, and vary the frequency until you find a **resonance**. With small motions of your hand near the right frequency, you should be able to make the middle of the rope move up and down with large amplitude. At the ends of the rope are **nodes**, points that never move. The middle of the rope where the amplitude is greatest is an **antinode**.

- The **standing wave** that you have created results from the interference of a wave traveling away from you and its reflection traveling back toward you. At the nodes, where the rope never moves, the two waves must have equal and opposite amplitudes. The amplitude in a wave reflected from a tied-down end is the negative of the amplitude of the incoming wave because this is the only reflected wave that can add to the incoming wave to produce a node where the medium is prevented from moving.

- Both ends of the rope must be nodes, so the length of the rope must be one-half wavelength, or an integer number of half wavelengths. Move your end of the rope at a higher frequency and see if you can find some other **modes** in which the standing wave has more than two nodes.

OVERVIEW *When either the source of a wave, or the detector, or both move through the wave medium, there are changes in the frequency detected. These can be understood by visualizing the wave pattern in the medium and the wavelength changes caused by the source motion. This visualization also explains what happens when the source speed is greater than the wave speed in the medium.*

A moving detector: A source at rest in a medium emits waves. They spread out in all directions. One crest has the shape of a spherical shell centered on the source. Its radius grows at the propagation speed of the wave in the medium. To be specific, let us consider sound. If you stand still and listen, crests arrive at your ear with exactly the frequency the source produces. But if instead of standing still you move through the air toward the source, the time between crests at your ear is shorter than the source period. If you move away from the source, the time between crests at your ear becomes longer than the source period. You hear a higher frequency when moving toward the source and a lower frequency when moving away from the source.

A moving source: If you stand still in air while a source of sound at a single frequency moves, you also hear shifted frequencies. Ahead of the source, in the direction toward which it is moving, the wavelength of the sound is shorter than if the source were at rest. The source emits a crest that moves away at the speed of sound and follows with its own velocity. When the first crest has moved one wavelength, the source has moved toward it and emits another crest, less than a wavelength behind. Because of the source's motion, the wavelength of the sound ahead of it is shorter than if it were at rest. Behind it, the wavelength becomes longer. When the sound reaches a stationary observer, it has a higher frequency if the source is approaching or a lower frequency if the source is receding.

- A nineteenth-century German physicist named Doppler discovered the frequency shifts we now call by his name.

KEY ILLUSTRATION 1

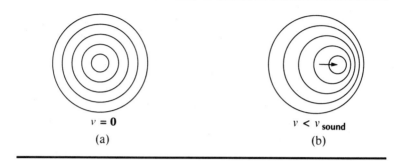

$v = 0$

(a)

$v < v_{sound}$

(b)

KEY ILLUSTRATION 2

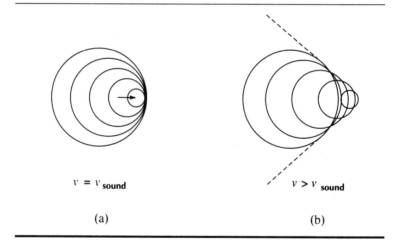

$v = v_{sound}$

(a)

$v > v_{sound}$

(b)

Supersonic source speeds: What happens if a sound source moves through the air with a speed faster than the speed of sound in air? At any point ahead of the source, it will arrive before the sound it emitted earlier arrives. Some time after it passes beside a detector, a high-intensity solitary wave arrives. This can be understood, like Doppler shifts, by visualizing the location in the air of crests emitted by the source. In illustration 1(b), the source moves slower than sound toward the right. The circles represent crests and are centered on the points from which they were emitted. The source in illustration 2(b) moves to the right faster than the speed of sound. The circles represent crests as above, but each is centered on a point outside the previous crest. There are lines (a cone in three dimensions) along which all crests align. This is the high-intensity solitary wave (sonic boom from a plane or bow wave from a speedboat) that forms behind a supersonic source. The cone becomes narrower as the source speed increases.

Key 50 Sound in air and in other media

OVERVIEW *The frequencies and wavelengths of sound in air, water, and other materials are given. The changes that occur as sound waves pass through a group of molecules are described.*

Frequencies, wavelengths, and speeds: Humans can hear sound waves with frequencies from about 20 Hz to about 20 kHz. Other animals can hear higher frequencies. Dogs can hear up to 30 kHz or 40 kHz, and bats use frequencies even higher to detect the structures in their environments.

- Sound travels at 330 m/s in dry air at 0°C. This speed increases by about 0.6 m/s for each 1°C increase in temperature and also increases with increasing humidity. Indoors, the speed of sound is around 340 m/s. In any gas, the speed of sound is about 80 percent of the average speed of the molecules, so the temperature dependence and its sign are not surprising. The sounds we hear have wavelengths in air of 1.7 cm (high frequencies) to 17 m—much smaller than our bodies to much larger.

- Sound travels 1 mi in air in about 5 s. Light travels much faster, so you can determine the distance to a thunderstorm by timing the interval between a lightning stroke and a thunder clap.

- In water, sound travels about four times as fast as it does in air; in steel, about 15 times as fast.

Compression and rarefaction: When a longitudinal (sound) wave passes through a gas, in addition to their random motions molecules have a tendency to move together in some places and move apart in others. The density and pressure of the gas change periodically by small amounts as the wave passes. Propagation of the sound wave occurs because the gas resists these changes. Regions of compression tend to expand, and regions of rarefaction tend to fill with molecules from nearby.

- Molecules pushing out of a compressed region produce compression in other regions; this is how the energy in a sound wave can get across a room after some periods of the wave.

Key 51 Reflection and refraction of sound

OVERVIEW *Sound waves reflect from walls and from other obstacles, just as a transverse wave in a rope reflects from a tied-down end. If the speeds of sound in two adjacent parcels of air differ, because their temperatures differ for example, then sound waves passing through them will bend.*

Reflection and reverberation: Reflected sounds—echos—arrive at a listener after a directly traveling sound wave. Long time delays give the impression of quiet repetitions of the original sound. More usual in typical rooms and halls, short time delays may produce interference as well as a slight prolongation of the sound. This reverberation is good; it makes music sound better and rooms feel more comfortable. Too much reverberation is bad and can produce harsh sounds. It is important to balance reflection and absorption of sound correctly to make an auditorium or concert hall acoustically pleasant.

Refraction: Variable winds and temperature differences often make the speed of sound vary from point to point in the air. If the speed of sound increases with height above the ground, the crests of sound waves travel farther up high than near the ground. The crests bend and continue to move perpendicularly to themselves. Sound energy that began traveling up can return to the ground. Sound from distant sources can be heard much better under these conditions—for example, across a lake at night when the water rapidly cools the air just above it.

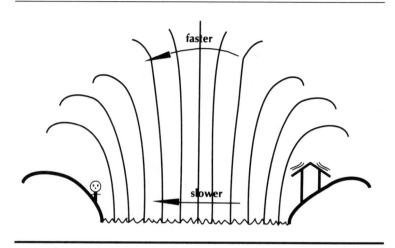

Pitch and frequency, loudness and amplitude, quality and waveform

---

OVERVIEW *The qualities of musical tones and the physical parameters of sound waves are related, sometimes simply, sometimes in a complex way.*

Pitch and frequency: The pitch of a musical note, high or low, is determined by the frequency of the sound wave that brings the music to you. A high frequency sounds like a high note; a low frequency sounds like a low note. The note A between middle C and high C is sound with a frequency of 440 Hz by convention, although variations of a few Hz may be found.

Loudness and amplitude: Loudness is the subjective impression that corresponds to the energy carried by a sound wave. A more energetic wave has a larger pressure amplitude and sounds louder.

- Our ears have a tremendous dynamic range; they can detect very low-energy waves and also deal with much higher energy waves without loss of loudness and frequency discrimination.

- The loudness of a sound is often measured using a quantity defined in terms of energy, but in such a way that its value corresponds, as well as we can tell, to the loudness we perceive. The unit used is the decibel, or dB.

- A decibel is 10^{-1} Bel. The Bel is a unit named for Alexander Bell, the inventor of the telephone.

- A sound level of 0 dB corresponds to the quietest sound we can detect. Ten dB has 10^1 times the energy; 20 dB has 10^2 times the energy; 30 dB has 10^3 times the energy and so on. Eighty-five dB is the level at which damage to the hearing apparatus may begin to occur. Loud amplified music, a gas-engined chain saw or weed whip, or a jet engine one plane-length away can be much louder than this.

Quality and waveform: A purely sinusoidal pressure variation with time at frequency f sounds to us like a tone with the pitch corresponding to f.

- Two sine waves with different frequencies may sound like two notes played together, or like just one note with a pitch corresponding to the lower frequency if the higher frequency is an integer multiple of the lower (880 Hz or 1320 Hz along with 440 Hz, for example).
- A set of waves with frequencies that are all integer multiples of the lowest, or fundamental, frequency produces the same pitch as the fundamental alone. But what a difference in quality! A single frequency produces a tone that sounds muffled, dull, and uninteresting. A fundamental mixed with many harmonics (higher integer multiples) produces a rich and interesting tone.

OVERVIEW *Sample questions of the type that might appear on homework assignments and tests are presented with answers.*

Wave motion:
- Sound waves in air travel with speeds v of about 340 m/s. What is the wavelength of a sound wave with a frequency of 300 Hz? The basic speed-frequency-wavelength relationship is $v = f\lambda$. Solving for wavelength $\lambda = v/f$ or, in this case, 1.13 m.

Standing waves:
- Standing waves of a violin string consist of two traveling waves, of equal amplitude, that reflect into each other at the ends. When the standing wave pattern on a particular string fits exactly one-half wavelength between the fixed ends, the frequency is 600 Hz. What is the frequency of the mode in which one full wavelength fits between the fixed ends? The wavelength of the new mode is half that of the 600 Hz mode. Wave speed on the string is the same for all waves, so the relation $v = f\lambda$ implies that the new frequency must be twice the old, 1200 Hz.

Doppler shifts:
- A horn emits a sound with a frequency of 340 Hz. The wavelength of the sound it emits from rest in still air is 1.0 m. If it moves through the air toward you with a speed of 34 m/s, then because of its motion how does the wavelength of the sound it emits toward you change? What is the frequency of the sound that you hear from the moving horn? The wavelength from the moving source is shorter in the region toward which it moves. The relationship $v_{wave} = f\lambda$ implies that the frequency you hear from an approaching horn must be higher than what you would hear from it at rest.
- If the horn of the previous question is emitting a sound at rest into still air, but you are moving away from the horn at 100 m/s through the air, then what is the speed of the wave from your point of view? What frequency do you hear in the situation of the previous question? You would experience a wave moving at 240 m/s with respect to yourself. The lower speed implies a longer time between arrivals of crests at your ears. Thus you would hear a lower frequency than you would hear at rest.

Theme 9 ELECTROSTATICS

A toms contain light, negatively charged electrons, and massive, positively charged nuclei. Electrical forces between charged particles hold them together. When two macroscopic objects exchange a large number of electrons and acquire net electrical charges, we can observe electrical forces. Observations first made on macroscopic objects contributed to our understanding of the structure of atoms and to the myriad uses we make of electricity.

Electrical forces and charges, Coulomb's law

OVERVIEW *Gravity is not the only fundamental (does that mean inscrutable?) force. Just as masses attract each other according to Newton's universal law of gravitation (Key 22), so do electrical charges attract or repel each other. Coulomb's law describes forces between two point charges. It is remarkably similar to Newton's universal law of gravitation.*

Charge: The source of the gravitational force is mass. Always positive, mass always produces attractive forces. The source of the electrical force is the electric charge, which can be positive (atomic nuclei) or negative (electrons). Charges never appear or disappear. All electrons have the same negative charge; all protons have the same equal and opposite positive charge. Hydrogen, which has one electron and one proton, acts as if it had no net charge. Large objects change their net charge by gaining or losing electrons. The presence of more electrons than protons produces a net negative charge. The presence of more protons than electrons produces a net positive charge.

Forces: Objects with net charges exert forces on each other.
- Two objects with like charges, either both positive or both negative, repel each other. In this respect, electricity is fundamentally different from gravity.
- Two objects with unlike charges, one positive and one negative, attract each other.
- The hydrogen atom, which has one electron and one proton, holds together because of the electrical attractive force between the electron and the proton.
- Why does hydrogen not collapse to an electron and a proton stuck tightly together? Why do the protons in a nucleus heavier than hydrogen not fly apart because they repel each other? Evidently there are other forces or rules other than Newton's at work (subject of later Keys).

Coulomb's law: Let $Q1$ and $Q2$ be two point charges separated by a distance d. The electrical force between them is

$$F = k \frac{Q_1 Q_2}{d^2}$$

and is attractive if the charges have opposite signs, repulsive if they have identical signs. This is Coulomb's law, named for the eighteenth-century French scientist Charles Coulomb. The l/d^2 dependence on distance is the same as the distance dependence in Newton's universal law of gravitation. As a consequence, the ratio of the electrical force to the gravitational force between two massive charged particles is the same, whatever their separation d.

Units:
- The SI unit of electrical charge is one Coulomb, 1 C, named for Charles Coulomb.
- The constant k in the force law is 9×10^9 Nm^{2}C^{-2}.
- The magnitude of the negative charge on one electron or of the positive charge on one proton is about 1.6×10^{-19} C.
- The values of the constant and magnitudes of charge reveal the fact that electricity began as an empirical science long before people were aware of the existence of electrons and protons.

Electricity and gravity: The electrical attractive force between an electron and a proton is 10^{40} times greater than the gravitational attractive force between the electron and the proton.
- Why are we usually unaware of electrical forces, while performing every motion we make with careful regard for gravitational forces? If electrical forces are 10^{40} times greater than gravitational forces, how is it that electrical forces are usually negligible in comparison to gravitational forces? In matter, one usually finds equal or nearly equal numbers of electrons and protons. Attractive and repulsive forces between large objects cancel, exactly or very closely. Gravitational forces between elementary particles add, never cancel, because all masses are positive. It is the very strength of the electrical forces that makes them so weak in everyday life; charged objects strongly tend to exchange electrons and so reduce their charges.
- Why do electricity and gravity have the same l/d^2 dependence on distance? Albert Einstein believed there must be a single theory of particle interactions that would include both electrical and gravitational forces, as well as any other fundamental forces. One universe, one theory—a grand goal that physicists still struggle to achieve.

Key 55 Conductors, insulators, semiconductors, and superconductors

OVERVIEW *Electrical charges pass easily through some materials but are immobile in others. Designers of electrical machines choose materials that allow them control over the flow of charges.*

Conductors: Metals are good conductors of electricity and heat for one reason. In metals, the most weakly attached one or few electrons from each atom move freely throughout the material, rather than remaining near "their" atoms forever.

- As electrons at one end of a wire move, the force they produce on electrons nearby changes, causing them to move, and the process propagates down the length of the wire quickly. You can add electrons at one end and take electrons out the other indefinitely.
- In water, many added materials exist as ions, with either positive or negative charges, that can move anywhere in the fluid. Salt dissolves and forms positive sodium ions (one electron missing) and negative chlorine ions (with one extra electron). Salt water is a conductor.

Insulators: In many nonmetals, electrons and ions are bound to fixed sites.

- Glass, plastic, and dry wood are common examples of insulators.
- Electrical charges usually do not move through insulators (exceptions include sparks and lightning).

Semiconductors: There are materials at the border between conductors and insulators. At very low temperatures they are insulators, but at room temperature a few electrons are thermally jarred loose from their fixed positions and can flow through the material. There are not as many free electrons as in a metal, so semiconductors are less effective conductors than metals.

- In semiconductor devices, small amounts of charge are moved around to control the flow of larger amounts of charge. Many such devices linked together form the heart of a stereo, television, computer, home theater, game box, music player, cellular phone, or other modern electronic gadget.
- Silicon is the most used semiconductor.
- Germanium, gallium arsenide, and a few others are also used as semiconductors in electronics.

Superconductors: In ordinary conductors there are frictional forces on moving charges that stop the motion in a very short time unless applied forces act to maintain it. Superconductors are materials that are relatively poor conductors at room temperature but become frictionless conductors at very low temperatures.

- Electric charges can flow in a closed superconducting circuit for years.
- Superconductors are used in electromagnets and in some very high-speed electronic devices.
- The theory of the behavior of superconductors is complicated because forces on charged particles that depend on the position and motion of many other particles must be included.

Key 56 Charging by rubbing, by contact, and by induction

OVERVIEW *How can one cause a few subatomic-sized charged particles to move from one macroscopic object to another? Often just rubbing two objects against each other is sufficient. Even without contact, a charged object's electrical forces cause charges in other objects brought close by to change their positions.*

Rubbing: Rubbing a piece of black, hard rubber with a piece of animal fur or with your own hair causes the electrons to move from the fur or hair to the rubber. Thus the rubber becomes negatively charged, and the fur becomes positively charged. This works well on very dry days but poorly when the air is humid. If you rub a piece of glass or transparent plastic with silk or wool, the glass or plastic becomes positively charged and the cloth becomes negatively charged.

Contact: A charged rubber, glass, or plastic rod is an insulator. Charges (either an extra electron or a missing electron) do not move around in the material but remain at fixed surface sites. These charges can be transferred to other objects, metal or insulator, by contact between the charge-bearing regions of the charged object and any part of an uncharged object. Electrical forces between the excess charges on the charged object tend to drive them apart and make some move to a neutral object in contact with the charged one.

Polarization: In a polarized conductor, electrons have moved to one side producing a negative charge there and leaving a positive charge behind on the other side. The conductor as a whole may be neutral, but its various parts are not. A neutral metal rod can be polarized by bringing a negatively charged piece of hard rubber close to, but not touching, one end. The negative charge on the rubber repels electrons in the metal. They move to the far end giving it a negative charge and leaving a positive charge on the end near the rubber.

Charging by induction: This is a way of charging a conductor such as an electroscope by connecting it to another larger conductor, polarizing the pair by bringing another charge near, and breaking the connection before removing the other charge. An electroscope is a vertical, metal shaft a few centimeters long, from which a flat plate and one very thin metal foil hang down. In some instances two thin foils hang without the plate (see Key Illustration). The foils repel and move apart when the electroscope is charged, or hang straight down if it is neutral. In charging by induction, the conductors are connected unpolarized (upper panel in Key Illustration) and disconnected when polarization has shifted some charge from one to the other (lower panel). The charge left on the smaller conductor by this process will be the **opposite** in sign from the charge that was brought near to polarize the connected pair.

KEY ILLUSTRATION

Charging an electroscope by induction

OVERVIEW *Like gravitational fields (Key 24), electric
fields help you visualize the size and direction of the force on
a small test charge as a function of its position. Inside a
closed, empty metal box, the electric field is zero everywhere, regardless of the charge distribution outside the box.*

Definition: The electric field **E** is defined in terms of the force **F** on a
test charge with a magnitude of q:

$$\mathbf{E} = \frac{\mathbf{F}}{q}$$

- The SI unit for electric fields is 1 N/C.
- The electric field in a region of space can be represented by drawing arrows at some points with lengths proportional to the magnitude of the field at those points and orientations giving the
  direction. This pattern of arrows can be replaced by a set of lines
  that are parallel to the arrows at every point. These lines of force
  show the field direction and indicate its magnitude by their density. Many lines close together represent a strong field.

The field of a point charge: The lines of force in the vicinity of a point
  charge are radially directed, pointing outward from a positive charge.
  To accurately represent $1/d^2$ dependence of the field on the distance
  from the charge, they must all terminate on the charge; none start or
  stop in empty space. The pattern is the same as the flow pattern of
  some conserved stuff flowing away from the charge. For this reason
  lines of force are sometimes called lines of electrical flux.

A shell of conductor: In a thin closed shell of metal, charge spreads out
  because like charges repel. A small test charge inside feels no force;
  forces from different parts of the surface cancel. The proof of this is
  the same as the proof given in Key 24 for gravitational forces.

- If a charge outside the shell is brought up close, it produces forces
  on the charges in the metal and they move, in such a way that the
  net electric field inside remains zero.
- Sensitive electronic circuits such as television receivers are
  enclosed in metal boxes to shield them from electronic noise.

Electrical potential energy
and electric potential

OVERVIEW *Just as raising a mass in a uniform gravitational field changes its total energy, moving a charge in an electric field changes its total energy.*

Electrical potential energy: The electric field magnitude is the force per unit charge that the field produces on a charge placed in it. If you move a charge Q a distance d in a field $\mathbf{E}$, in a direction opposite to that of the field, you must do work EQd. The electrical potential energy of the charge has increased by this amount as a result of its change of position.

Electrical potential: In the example above, the change in potential energy per unit charge is just Ed. This is the change in the electric potential between the two points separated by d. The electric potential of a system can be defined at all points in space; it is a field like $\mathbf{E}$, but it has scalar rather than vector values. The SI unit of electric potential is the volt (V):

$$1\ V = 1\ \frac{J}{C}$$

- The electric field has two equivalent units:

$$1\ \frac{V}{m} = 1\ \frac{N}{C}$$

- The volt is named for Allesandro Volta, an Italian scientist who was a pioneer in the study of electricity.
- Electric potential is sometimes referred to as voltage.
- Flashlights use batteries rated at 1.5 V, and automobiles use 12 V batteries. These are devices, through which charge can flow, that give each C of charge flowing through an increase in potential energy of either 1.5 J or 12 J.
- The two terminals of a battery have electric potentials that differ by the battery voltage.

OVERVIEW *Sample questions of the type that might appear on homework assignments and tests, with answers.*

Electric forces and fields:

- What is the electric field E at a distance of $d = 4$ cm from a positive charge of $Q = 10^{-4}$ C on the tip of a glass rod? Coulomb's law for the magnitude of the field is $E = kQ/d^2$ with $k = 9 \times 10^9$ Nm^2C^{-2}, so $E = 9 \times 10^9 \times 10^{-4}/(0.04$ m$)^2$ or 5.6×10^8 N/C, which is 5.6×10^8 V/m. The direction of the electric field is away from the tip of the glass rod.

- In electric field E of 300 V/m = 300 N/C, what is the force F on a charge of 10^{-4} C? $F = QE$, which is 300 N/C $\times 10^{-4}$ C or 0.03 N.

- The electric field E 1 m away from a point charge has magnitude 24,300 V/m. What is its magnitude 3 m away? What does the electric field E 1 m away from the source become when every charge in the system is made $\frac{1}{3}$ as large? Because $E \propto 1/d^2$, the field 3 m away is $\frac{1}{9}$ as large as the field 1 m away, 2700 V/m. If the charge is reduced to $\frac{1}{3}$ of its original value, the field everywhere becomes $\frac{1}{3}$ as large: 8100 V/m at 1 m and 900 V/m at 3 m.

Electric potential:

- Two metal plates are separated by 3 cm and have length and width much greater than their separation. The potential difference between them is 30,000 V. What is the magnitude of the electric field E halfway between the two plates? What is the magnitude of the electric field E, not between the two plates of the previous question, but outside of them? Between the plates $\Delta V = Ed$ so $E = 30,000$ V/0.03 m or 10^6 V/m. Outside the region between the plates the field is much smaller.

Units of electrical energy:

- An electron-volt (eV) is a convenient unit of energy in discussions of one or just a few charged particles. Electric utilities measure energy in kilowatt-hours (kWh). How many eV are there in 1 kWh? The charge e on one electron has a magnitude 1.6×10^{-19} C, so 1 eV = 1.6×10^{-19} CV = 1.6×10^{-19} J. 1 J = 1 eV/(1.6×10^{-19}). 1 kWh is 10^3 Js^{-1} times the number of seconds in one hour, 3,600 s, or 3.6×10^6 J. Multiplying by the number of eV in 1 J, 1 kWh = 1 eV $\times (3.6 \times 10^6)/(1.6 \times 10^{-19})$ or 1 kWh = 2.3×10^{25} eV.

This Theme describes the way that electrical charges flow in circuits formed from conducting material. Mastery of these circuit techniques has given us the use of electrical power for lighting and for mechanical work, electronic devices for reproduction of pictures and sound, and computers that control these devices and perform many other tasks. The principles of current electricity are presented here.

Electric current, resistance, and Ohm's law

OVERVIEW *For a current of electrical charge to flow in a conductor despite frictional forces, there must be an energy source driving the flow. Just as water will flow downhill as long as it is pumped up and released, electric charge flows as long as a circuit element increases its potential energy and then releases it. In ordinary conductors, the motion of electrons is opposed by frictional forces. Ohm's law describes how the balance between driving forces in sources of potential difference and frictional forces determines the current that flows.*

Flowing charge: Electric current is flowing electric charge. In currents in wires, electrons move through the material, just as water molecules move through a pipe in which water flows.

Units and examples:
- If 1 C of charge moves past a point in a time of 1 s, the magnitude of the current is 1 A, one ampere or simply one amp.
- The ampere is named for the early nineteenth-century French scientist André Marie Ampere, who studied the magnetic effects of electric currents.
- Currents in incandescent light bulbs are about 1 A; in a toaster, about 10 A; and in smaller electronic devices, typically 1 mA.
- Even though negatively charged electrons are the particles whose motion produces a current, by convention we take the direction of current flow to be the way positive charges would move if they were carriers.

Voltage and current: Water in a long horizontal pipe does not flow by itself. If the pipe rests on a hillside so that water has a higher potential at one end than at the other end, then the water in the pipe will flow. Current flows **through** a wire or other circuit element **across** which there is an electric potential difference.

Resistance and Ohm's law: As more and more current flows in a wire, the frictional force opposing the current grows. The driving force from a source of potential difference has a fixed magnitude, so the current grows only until the frictional forces become equal to the driving force. In many materials and electrical components (called linear or Ohmic), but not in all, the current flowing is proportional to the potential difference across the component. This is Ohm's law:

$$I = \frac{V}{R}$$

I is the current, V is the potential difference, and R is called the resistance of the component.

- The resistance R has units of 1 V/A = 1 Ω, the upper-case Greek letter omega. The unit Ω is called an ohm. G.S. Ohm, a German physicist, discovered his law in the early nineteenth century.

Key 61 DC and AC electrical power

OVERVIEW *Direct currents (DC) are constant in time—10 A to the right, for example. Alternating currents (AC) flow in household wiring. The magnitude of an AC current is a sine or cosine function of time, like the displacement of a pendulum. The power extended by a voltage source pushing a current against other forces is just the mechanical work per time done by the electrical forces on the charged carriers of the current.*

AC frequency: In the United States, the frequency of household voltage and current is 60 Hz. In an incandescent lamp connected to an AC outlet, the potential difference across a fine wire (filament) grows to more and more positive values, and so does the current through it. Then both decrease, pass through zero, and take on negative values. Friction heats the filament whichever way the current flows, so it becomes hot and glows just as it would with DC current through it.

DC power: For a component with current I through and voltage V across, the electrical power delivered is:

$$P = IV = I^2R = \frac{V^2}{R}$$

The first equation is always useful. The last two apply only to an Ohmic component with resistance R.

AC power: An AC voltage that delivers to a resistor the same average power over time as a DC voltage of 117 V is called 117 V AC. Its maximum amplitude must be greater than 117 V, because at some times its amplitude is much less than the maximum. Likewise, an AC current that delivers to a resistor the same time-averaged power as a 0.5 A DC current is said to be 0.5 A AC.
- Household wiring is 117 V AC, delivers 0.5 A AC to a light bulb with a resistance of 234 Ω, and so delivers a time-averaged power of 59 W.

Kilowatt-hours: Household energy use is measured by electric utilities that charge us a fixed price per energy unit, usually something like $0.12 kWh (kilowatt-hour; see Key 13). At the rate just quoted, a 60 W light bulb can be powered for about 1.4 hours for a penny.

Key 62 Series and parallel circuits

OVERVIEW *Circuits can be arranged so that the current from the voltage source flows through a number of components in turn, or so that it divides and some flows through each of a number of components. With Ohm's law, conservation of charge, and the fact that electric potential has one value at each point in space, one can completely analyze even complicated circuits.*

Parallel circuits: In homes and automobiles, lamps and appliances are connected across the voltage source in parallel. Each component with two terminals is connected by two wires to the voltage source, component terminal to source terminal. Each component has the full source voltage across it, and Ohmic components have currents through them that do not depend on what other components are connected. One burned-out lamp has no effect on the operation of other lamps and appliances. The total current the source delivers is the sum of the currents drawn by each component connected. High-resistance components have small currents, and low-resistance components have larger currents.

Series circuits: In a series circuit there is only one path for current to follow. After leaving the source, current enters the first component, leaves and enters the second, and so on until it flows back into the source. Like the string through a string of beads, it passes through each one in turn. The total current from the source is the same as the current in any component. The voltage across the source is the sum of the voltages across each component. High-resistance components have high voltages across them, and low-resistance components have lower voltages across them.

Computations:
- **Kirchoff's laws** along with Ohm's law give us what we need to compute currents and voltages in circuits. Kirchoff's laws are, first, that the sum of all currents flowing into a point in a circuit is zero (conservation of charge), and second, that the sum of all potential changes around a closed path in a circuit is zero (electric potential has one value at each point in space). It follows that, in

a parallel circuit of Ohmic components, all resistances, $R_1, R_2, \ldots,$ R_N could be replaced by one R_{eff} with a resistance

$$\frac{1}{R_{eff}} = \frac{1}{R_1} + \frac{1}{R_2} + \ldots + \frac{1}{R_N}$$

and the voltage source would have to deliver the same current to it.

KEY ILLUSTRATION

$R_1 \qquad R_2 \qquad R_3 \qquad R_4$

- In a series circuit of Ohmic components, all the individual resistors could be replaced by one R_{eff}

$$R_{eff} = R_1 + R_2 + \ldots + R_N$$

KEY ILLUSTRATION

$R_1 \qquad\qquad R_2 \qquad\qquad R_3 \qquad\qquad R_4$

Key 63 Key questions with answers

OVERVIEW *Sample questions of the type that might appear on homework assignments and tests are presented with answers.*

Ohm's law:
- A 103 Ω resistor is connected across a 12 V battery. What current flows? What power is dissipated in the resistor? The current is $I = V/R$ or 12 V/10^3 Ω, which is 0.012 A or 12 mA. The power V^2/R is $(12 V)^2/10^3$ Ω or 0.144 W, 144 mW.

Series and parallel resistances:
- What single resistance acts in electrical circuits exactly like a 400 Ω resistor and a 600 Ω resistor connected in series? When these two resistors are connected in series across the terminals of a battery, in which one is more electrical energy converted to heat? The equivalent single resistor has a resistance that is the sum of those of all the series-connected components, 10^3 Ω. In a series circuit, the current in each resistor is the same, so the best expression for computing power is $P = I^2R$. This shows that the greater power is dissipated in the component with the larger R, the 600 Ω resistor in this case.

- What single resistance acts in electrical circuits exactly like a 200 Ω resistor and a 400 Ω resistor connected in parallel? The equivalent single resistor has a resistance that is the inverse of the sum of the inverses of the resistances of all the parallel-connected components, $1/R_{eq} = 1/R_1 + 1/R_2$. In this case, $1/R_{eq} = (1/200 + 1/400)$Ω^{-1} or $(3/400)$Ω^{-1} so $R_{eq} = (400/3)$Ω or 133 Ω. In a parallel circuit the potential difference across each resistor is the same, so the best expression for computing power is $P = V^2/R$. This shows that the greater power is dissipated in the component with the smaller R, the 200 Ω resistor in this case.

Magnetic forces, observed many centuries ago as were electrical forces, are related to electrical forces, a fact realized only about two centuries ago. Magnetic forces are as central to the operation of electrical and electronic machines as electric forces.

OVERVIEW *Magnetic forces appear, in addition to electric forces, when electric charges are moving. But magnetic forces were first observed acting between magnetic materials that were not moving.*

Magnetic forces: Compass needles, bar magnets, and the earth itself exert magnetic forces. Like electric forces, magnetic forces may be attractive or repulsive and become weaker as the distance between magnets increases.

Magnetic poles: Long, thin magnets, like compass needles, behave as if the sources of their magnetic forces were concentrated near the ends. If the end of another magnet attracts one end, it repels the other end. These sources are called **poles** that even in experiments involving the testing of many magnets in turn, obey a simple rule: **like poles repel; unlike poles attract.** On the earth, with a compass in hand, you can identify any pole of any magnet as north-seeking or south-seeking. (What kind of pole, then, lies beneath the earth's surface near its north pole? It is a south-seeking pole).

  • Poles cannot be isolated. If you break a long bar magnet anywhere between its two unlike poles, the resulting piece will have two unlike poles. It is as if in electricity, you can polarize materials but never separate positive and negative charges very far from each other. Currently, elementary particle physicists are trying to understand why we observe the particles and forces we do, rather than others. For these physicists, the absence of magnetic monopoles remains a challenging fact.

Forces between poles: The magnetic force between two poles is like the electric force between two charges:

$$F = \frac{AP_1P_2}{d^2}$$

P_1 ad P_2 are pole strengths, d is their separation, and A is a constant of proportionality.

  • **F** is attractive between unlike poles, repulsive between like poles, and always acts along the line between them. This simple force cannot be observed experimentally, because poles cannot be isolated.

OVERVIEW *Like electric and gravitational fields, magnetic fields can be used to visualize the pattern of forces on a test pole in any system of fixed poles. Because the force between poles has a $1/d^2$ dependence on their separation d, lines of magnetic field spread out from poles like a flow of some conserved stuff.*

Magnetic fields: The magnetic field near a pole looks like the electric field near a charge. Its magnitude at every point is the ratio of the force on a small test pole at that point to the strength of the test pole, and its direction is the direction of the force on a positive test pole. Because there are no isolated poles, this definition is not often used; indeed, there is no firmly established convention declaring which kind of pole is to be treated as positive.

• The earth's magnetic field in a room is uniform and can be represented by equally spaced, parallel lines. The force on a test pole has the same magnitude and direction everywhere in the room. A pair of equal and opposite poles, for example, those in a compass needle, feel zero net force but are acted on by a torque that tends to align them so their separation is parallel to the field direction.

Sources of magnetic fields: An electric charge moving in a magnetic field is acted on by a force that does not act on a charge at rest; it is proportional to the charge's speed. The source of this force, inside the magnetic material that acts as if it contained a pole, is also moving electric charge. This idea was first proposed in the 1820s by André Marie Ampere, a French physicist. Electrons, which are spinning on their own axes and circulating around atomic nuclei, produce magnetic fields. In many materials, the fields from all the electrons in one atom cancel each other. In others, fields from different atoms have different orientations and tend to cancel. In materials we perceive as magnetically active, the fields from many electrons are adding without cancellation.

Magnetic flux: Lines of magnetic field or magnetic flux do not start or end in empty space, but rather look like lines of flow of some conserved stuff. In the picture of magnetism based on poles, the lines start and end on poles. But there are no poles, only circulating charges—loops of electric current. Lines of magnetic flux are closed loops that link the current loops that produce them like links of a chain or of a bracelet.

KEY ILLUSTRATION

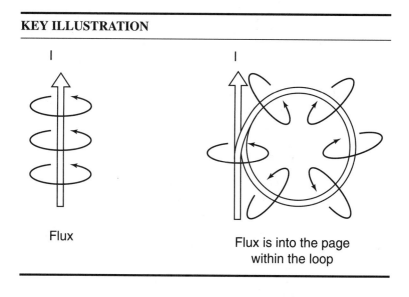

Flux

Flux is into the page
within the loop

Key 66 Magnetic domains

OVERVIEW *In strong magnets, fields from the electrons in one atom do not cancel, and fields from atoms in one region add together without cancellation.*

Domains: Atoms that produce nonzero magnetic fields produce magnetic forces on each other. In solids, these forces do not change atomic positions, but they can align the moving charges in atoms. The forces tend to align individual atomic magnets the same way in some materials, the opposite way in others. In those tending to align the same way, magnetic domains form.

- All the atomic magnets inside a domain are aligned in the same direction; their magnetic fields are likewise aligned and add together to produce strengths we can detect in simple experiments with magnetic materials.
- Magnetic domains are typically smaller than 1 mm.

Magnetization: Even in a strong magnetic material—a **ferromagnetic** material—the domains may be randomly oriented or aligned more in one direction than any other. The material may be magnetized or not.

- The term *ferromagnet* means a magnet like iron (Fe) and comes from the latin name for iron, *ferrum.*
- Soft iron, which is not cut or bent after cooling from high temperature, responds strongly to another magnet but is not itself a magnet. If a very strong magnetic field is applied and then removed, the iron is left magnetized. Its domains, which are randomly oriented initially, grow, shrink, and change orientation in response to applied magnetic forces.
- Frictional forces oppose the changes in orientations of domains. You can make a piece of iron magnetize more strongly by hitting it with a hammer while it is exposed to a strong applied field. The blows from the hammer jar domain walls loose from the frictional forces that restrain them.
- Hard or permanent magnets have strong enough frictional forces to preserve their domain structure at room temperature in external fields that are not too strong.
- Soft magnets have weak frictional forces and respond to changing fields by changing their domain structure (i.e., their magnetization). Soft magnets are useful in magnetic tape and computer disk recording and playback heads.

Key 67 Electromagnets and typical magnetic field strengths

OVERVIEW *Just as moving charges in atoms produce magnetic fields, moving charges in electrical circuits also produce magnetic fields.*

Air-core electromagnets: A loop of wire carrying an electric current produces a magnetic field that has the same shape as the field of a magnet with poles separated by a line parallel to the loop's axis. This is a dipole (two-pole) field (Key 65). To make a strong electromagnet, coil a long wire into many loops so that the current passes through each one in turn. The fields from the loops add together. This is called a solenoid.

Superconducting solenoids: In an electromagnet higher currents produce higher fields. Ohmic losses ($P = I^2R$ rate of heating) limit the currents that can be used. In superconductors there are no Ohmic losses. High constant magnetic fields are often produced by air-core superconducting solenoids that use power only to start up and cool the superconducting material.

Iron-core electromagnets: A soft ferromagnet placed within a solenoid magnetizes when current is driven through the solenoid. Its field adds to that of the solenoid. This method of increasing the field produced by a given current is efficient and useful up to field strengths at which all the domains in the ferromagnet are aligned.

Strengths of magnetic fields: The SI unit of magnetic field strength is T, or tesla, named for N. Tesla, who designed practical AC motors and generators that allowed modern electrical utilities to develop.
- The strength of the earth's magnetic field at its surface is about 5×10^{-5} T.
- Permanent magnets produce fields of about 10^{-2} T near their surfaces.
- Air-core solenoids of ordinary conductors produce fields inside of up to 10^{-1} T; more only if special cooling is provided.
- Iron-core solenoids produce fields approaching 1 T in their gaps.
- Superconducting air-core solenoids provide fields up to 21 T for research applications.
- Superconducting solenoid: Normal solenoid hybrid electromagnets can produce constant fields up to 45 T.

Key 68 Magnetic forces on moving charged particles and on current-carrying wires

OVERVIEW *Electric currents produce magnetic fields and are also acted on by magnetic forces when they flow in a region of space with a nonzero magnetic field.*

Moving charge: A charge at rest in a magnetic field feels no magnetic force. A charge moving parallel to a magnetic field also feels no force. A charge moving perpendicular to a magnetic field feels a force that is proportional to its charge, its speed, and the strength of the field. The direction of the force is perpendicular to both the velocity and the field.

- Q coulombs of charge moving with speed v through magnetic field with strength B feel a force:

$$F = BQv$$

The force is in N if the speed is in ms^{-1}, the charge is in C, and the field strength is in T. If **B** and **v** are not perpendicular but have angle θ between them, then the force is $BQ\sin\theta$, proportional to the product of B and the component of v perpendicular to it or of v and the component of B perpendicular to it.

Currents in wires: A current-carrying wire contains electrons that cannot escape but are held inside by forces that act near the surfaces. The electrons feel the same magnetic forces that particles moving in a vacuum feel, when a magnetic field is applied. The force is perpendicular to the field and to the wire and has magnitude:

$$F = BIL$$

B is the field strength in T, I the current in A, and L the length in m of the section of wire on which the force F acts. The formula tells the force per length of wire in the field. If the wire is not perpendicular to the field, the part of the field that is perpendicular to it must be used: $BIL\sin\theta$ if θ is the angle between the wire and the field. The force is perpendicular to both the field and the wire.

Directions of forces: Consult your textbook for right-hand rules that you can use to find the directions of the forces.

Forces between currents: Two parallel wires close to each other that are carrying currents attract or repel. The current in one produces a perpendicular field at the other. Currents in the same direction attract; those in opposite directions repel.

Key 69 Electrical meters and motors

OVERVIEW   *The forces on current-carrying wires are used to make meters that indicate the size of a current and motors that convert electrical energy to mechanical work.*

Meters: Place a compass just above a wire that is oriented in the north-south direction. With no current, the compass points north. As the current increases, the compass needle swings closer to an east-west orientation. First observed in 1820 by Hans Christian Oersted, a Danish physicist, this established the relation between electricity and magnetism. A physically similar but mechanically more sophisticated, better engineered device for measuring current is the **galvanometer**, in which the current passes through a coil that is free to rotate in the field of a permanent magnet. A spring provides a restoring torque proportional to the amount of rotation away from a zero-current position, and an attached pointer moves over a scale as the coil rotates. Typically full-scale deflection, 90° rotation, is produced by a current of 5×10^{-5} A.

- A voltmeter is a galvanometer with a large resistance in series: an ammeter a galvanometer with a small resistance in parallel. Knowing the resistance of the galvanometer, its full-scale current, and the rules for series and parallel circuits (Key 62), you can determine the resistors needed to make a voltmeter or ammeter with a specified full-scale reading.

Motors: A motor has the same configuration as the galvanometer just described; however, the motor doesn't have the pointer or any restoring spring. In a strong, stationary magnetic field provided by a permanent magnet or by an electromagnet, a coil of heavy wire can carry a large current. Free to rotate on bearings around an axis in the plane of the coil and perpendicular to the fixed field, this electromagnet aligns like a compass needle in the fixed field when a current flows. Change the current direction in the rotating coil and it tends to align 180° away from its first orientation. AC current fed through sliding contacts mounted on the axle produces continuous rotation at the same frequency as the AC current frequency.

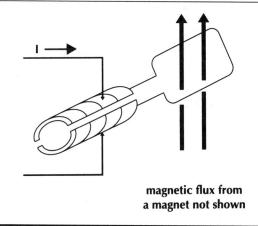

**magnetic flux from
a magnet not shown**

- A DC motor requires a **commutator**, which is a switch controlled by the shaft angle of the motor that changes the direction of the current in the rotating coil. A split-ring commutator is no more complicated than the sliding contacts needed in an AC motor. Many motors with commutators are "universal," i.e., designed to operate on either AC or DC current.

Key 70 The earth's magnetic field

OVERVIEW *The magnetic field of the earth that aligns a compass needle is produced by electric currents in the liquid core of the earth.*

We live on a magnet: The magnetic field of the earth is a dipole field, just like that of a bar magnet.
- The south-seeking pole, which attracts north-seeking poles of other magnets, is far below the surface of the earth under the Arctic Ocean, at about 82° north, 110° west, and moving north at about one half degree per year.
- The north-seeking pole is under the ocean between Australia and Antarctica.
- Away from the poles, the field at the surface indicates direction, with some error because the magnetic poles are not directly under the rotational poles. The deviations are well known and are described on most maps.
- The earth's field has a vertical component as well as the horizontal one we use to find the north-south direction. Zero at the equator, it grows at higher latitudes. It could be used to measure latitude, but the elevation of the sun and stars has proven more reliable.
- Refer to your textbook or another general physics text for a diagram of the earth's magnetic field.

The dynamo: Dynamo is a name used for electrical generators that convert mechanical work to electrical energy. It is especially suitable for the process in the earth that generates its magnetic field, because that process is not stationary.
- Geological evidence, the direction of magnetization in magnetic rocks, shows that the field has reversed many times, and has been "off" or at least much smaller that its current value of about 10^{-4} T for extended periods in the past.
- The process in the flowing, molten, conductive liquid core of the earth that produces the field is not fully understood.

OVERVIEW *Sample questions of the type that might appear on homework assignments and tests are presented with answers.*

Magnetic poles: Which pole of a strong bar magnet will attract the north pole of another magnet? Which pole will attract an iron nail? Which pole will attract a glass rod? Like poles repel; unlike poles attract. The south pole of a strong magnet will attract the north pole of another magnet. Either north or south poles will attract an iron nail by inducing the opposite pole in the nearest part of the nail. Neither pole will attract a glass rod; glass has no strong magnetism.

Magnetic forces on moving charges:

- Consider a horseshoe magnet and a current-carrying wire passing between its poles. If the force between them is 1 N when the current in the wire is 2 A, what does the force become when the current is increased to 8 A? The force is proportional to the current and so becomes four times greater—4 N.

- A magnetic field points straight down. An electron moves through it at 10^7 m/s, east. In what direction is the force on the electron? The force is perpendicular both to the velocity and to the field and so must be either north or south.

- A magnetic field points west. An electron moves through it at 107 m/s, east. In what direction is the force on the electron? When the velocity and the field are parallel, the force is zero.

- Consider a constant magnetic field at rest in your laboratory. Can it change the speed of a charged particle that moves through it in a direction NOT parallel to the field? Can it change the direction of the velocity of a charged particle that moves through it in a direction NOT parallel to the field? There is a force on the moving charged particle, because its velocity is not parallel to the field. The force is always perpendicular to the velocity, and so it never does any work on the particle. It can change the direction of motion but not the kinetic energy; the speed remains constant.

ELECTROMAGNETIC
INDUCTION

This Theme explores the electrical forces and currents in conductors that are produced by magnetic fields that change in time or by motion of the conductors through magnetic fields. Practical applications of electric generators and transformers are described, and the stage is set for the subject of the next two Themes—electromagnetic radiation.

INDIVIDUAL KEYS IN THIS THEME

72   Faraday's law of electromagnetic induction

73   Electric generators, AC and DC

74   Transformers

75   Self induction and Lenz's law

76   Key questions with answers

OVERVIEW *Charges moving in a magnetic field experience a force. This is just as true of motion produced by moving a wire through the field as it is of motion associated with a current flowing in the wire. Moving a wire through a magnetic field tends to induce a current in the wire.*

Reciprocity: If current in a wire deflects a compass needle, should not applying a magnetic field to a wire produce a current? Familiarity with Newton's third law of motion inclines one to believe in this simple idea: If one pushes down on part A of a complicated system and observes that part B moves left, then surely pushing B right will make A move up. This is reciprocity. Often it can be proven by a chain of third-law arguments (Key 9). A scientist without the understanding needed to construct such a chain may find a guessed-at reciprocity setting the direction of his or her research.

Faraday and Henry: In the early 1830s, Michael Faraday in England and Joseph Henry in Princeton, New Jersey, observed that magnetism produced electricity. Moving a wire in a magnetic field produces a current in the wire, if it is part of a closed circuit, or a voltage between the ends if it is not. The voltage is called an induced EMF (electromotive force) to emphasize that it can make charges move even though it is not a Coulomb force. Faraday published these observations first in a widely read scientific journal of the time.

Faraday's law: In a modern demonstration, a coil of many turns is connected to a galvanometer (Key 69). One end of a bar magnet is rapidly pushed into the coil. While the magnet is moving, the meter deflects to one side. When the magnet is not moving, the meter reads zero. While the magnet is rapidly removed, the meter deflects to the other side. Faraday's law states that the induced voltage (EMF) in a coil is proportional to the product of the number of loops in the coil times the rate of change of the magnetic field linking one loop. In a closed circuit, the induced current depends on the induced voltage and other circuit parameters.

  • The changing magnetic field in Faraday's law may arise from a moving magnet near the coil, from motion of the coil in a fixed magnetic field, or by a change in the current in a nearby electromagnet.

OVERVIEW *By Faraday's and Henry's induction, making coils rotate in magnetic fields can produce electrical power.*

AC generators: Every electric motor (Key 69) is also an electric generator. If a coil is made to rotate in a constant magnetic field that is perpendicular to its rotation axis, then the voltage between the ends of the coil changes with time as a sine function of time. Electrical connections to the coil ends by sliding contacts (Key 69) can be used to produce an AC voltage between a pair of wires or an AC current through a load connected between the wires.

DC generators: A generator with a commutator instead of the simplest sliding contacts (see the figure in Key 69) produces a voltage that varies with time but always has the same sign. The voltage and the current that it produces change their magnitude with passing time, but not their sign.

- T. Edison, N. Tesla, and others built useful electric power systems within 60 years of the discoveries of Faraday and Henry. Their generators, turned by mechanical work, produced enough electrical energy to illuminate rooms with electric lamps and power motors that deliver useful mechanical work (Key 69). Edison used DC voltage and current and, because of power losses in the resistances of the wires used to distribute the power (Key 74), could only serve businesses within a few blocks of his generators. Tesla used AC voltage and current and transformers to serve wider areas (Key 74).

Key 74 Transformers

OVERVIEW *The voltage (maximum potential difference between the wires) from a DC generator is fixed; increasing the angular speed can increase it, but at fixed speed it has fixed value. AC voltage can be changed with negligible loss of power, thanks to Faraday and Henry.*

Induction: Consider two coils of wire close to each other and with a common axis. If a current in one coil produces a magnetic flux, most of the flux passes through the other coil. To keep our considerations simple, we shall assume all the flux from one coil links the other.

- The flux produced by a current in one coil, let us call it the primary coil, is proportional to the product of the current in the primary times the number of turns in the primary.
- The voltage induced in the secondary is equal to the product of the rate of change of that flux times the number of turns in the secondary.
- The voltage induced in the *primary* by the changing flux is the product of the rate of change of the flux times the number of turns in the primary. So, the ratio of the potential difference across the secondary to that across the primary is equal to the ratio of the number of turns in the secondary to the number of turns in the primary.

$$\frac{V_p}{n_p} = \frac{V_s}{n_s}$$

- With a secondary that has more turns than the primary, you can make a small voltage into a bigger one.

Energy conservation: You can use a transformer to make a bigger AC voltage, but you cannot avoid the law of energy conservation. The power coming out of the secondary is, at best, equal to that going into the primary.

- Electric power is the product of voltage times current; if the secondary voltage is bigger, and the power is conserved, then the secondary current is smaller than the primary current. If there are no losses in the transformer, then

$$V_p I_p = V_s I_s$$
$$I_s / I_p = N_p / N_s$$

- You can use a transformer to multiple voltage, or to multiply current, but energy conservation requires that their product be constant, or decrease a bit if there are losses.
- Connect a light bulb with resistance R across the secondary (N_s turns) of a transformer. The primary of the transformer (N_p turns) looks like a resistor R_p to a power source driving it. What is R_p? Knowing

$$R = \frac{V_s}{I_s}$$

and

$$R_p = \frac{V_p}{I_p}$$

and using the formulae in the last paragraph, you can show that

$$R_p = R\frac{N_p^{\,2}}{N_s^{\,2}}$$

- A transformer that steps up voltage looks, to a power source, like a smaller resistor than the one actually connected across its secondary. A transformer that steps down voltage looks, to its power source, like a bigger resistor than the one connected across its secondary.

Electrical power transmission: Let us consider two fat copper wires used to transmit electrical power from a generator in the country to a **load**, a city full of light bulbs, motors, stereos, televisions, and computers. With a cross-sectional area of 2 cm^2 and a combined length of 10^5 m, the wires have a resistance of 10 Ω. To deliver 10^9 W of power at 117 V potential difference **in the city, without using transformers,** the wires would have to carry a current of almost 10^7 A. The voltage between the wires at the generator would have to be almost 10^8 V (Ohm's law) and the power lost in the wires by Ohmic heating would be 7.3×10^{14} W, much greater than the power delivered to the city. These numbers are impossibly large.

- Transformers are used in electrical power transmission to produce a very large potential difference, 10^5 V to 10^6 V, between the wires. To carry 10^9 W of power at 10^6 V potential difference, the wires must carry 10^3 A of current. The potential differences along the wires add to only 10^4 V (Ohm's law), a small fraction of the 10^6 V between them. The power lost in heating the wires is 10^7 W, 1 percent of the power they transmit. Transformers in the city reduce the voltage to usable values.

OVERVIEW *If you change the current through a coil, the magnetic flux linking the coil changes, and this produces a voltage between the ends of the coil. To understand inductors, automobile ignition coils, or motors, you must be aware of and use this fact. Lenz, a late nineteenth-century physicist, measured the signs of induced voltages in a variety of situations and summarized his results in a way that pleases fans of reciprocity (Key 72) and the third law (Key 9).*

Induced current: The current in a coil produces a magnetic field that links the coil. Changing the current changes the field. But a changing field means that the flux linking the coil is changing. That tends to induce a current in a closed circuit, in addition to what is already there in the coil because of external sources (Key 74).

Induced voltage: A voltage appears across the coil when the current in it is changing, with a sign and a magnitude that tend to produce the induced current. This is the induced electromotive force (EMF), a fancy name for a voltage that arises from nonelectrostatic sources. This voltage is equal in magnitude to the rate of change of magnetic flux linking one turn of the coil, multiplied by the number of turns linked.

Lenz's law: The modern statement of Lenz's law is this: an induced voltage tends to oppose the change that is inducing it. Consider a coil connected to a battery, so a constant current is flowing through it. If you break the connection by removing a wire from a terminal, you will see a spark jump between the end of the wire and the terminal. Remove the wire slowly and the spark may persist for a long time. As the current and the flux begin to decrease, an induced voltage grows, with a sign that opposes the decrease in the flux. If the current had been flowing from wire to terminal, then Lenz said that the induced voltage will make the wire positive relative to the terminal, tending to pass the previously flowing current through the spark.

OVERVIEW *Sample questions of the type that might appear on homework assignments and tests are presented with answers.*

Generators:

- Consider the motor in Key 69 operated as a generator. Is the current that flows in the circuit (not completely shown on the left) AC or DC? The current in the coil on the right is AC. The commutator at the center of the figure switches connection between the coil and the external circuit on the left, so the current on the left always flows in the same direction; it is DC.

Induction:

- A coil of 100 turns of wire encloses an area of 0.2 m². A magnetic field, **parallel to the plane** of the coil, is increasing at a rate of 1 T/s. Is there a potential difference between the ends of the wire? No lines of magnetic field ever link the coil, so there is no potential difference.
- A coil of 100 turns of wire encloses an area of 0.2 m². A magnetic field, **perpendicular to the plane** of the coil, is increasing at a rate of 1 T/s. Is there a potential difference between the ends of the wire? The number of magnetic field lines linking the coil grows steadily, so there is a potential difference.

Transformers:

- An AC transformer has a primary winding with 100,000 turns and a secondary winding with 1000 turns. An AC voltage of 1000 V is applied across the primary. What is the voltage across the secondary? A resistor connected across the secondary draws a current of 100 A. What is its resistance? What is the current flowing in the primary circuit? What resistor, connected across 1000 V, would draw the same current as the transformer of the previous question? The voltage across the secondary is $V_s = V_p \times (N_s/N_p)$, which is 10 V. The resistor that draws 100 A with 10 V across it has a resistance of 0.1 Ω. The power flowing into the primary must be equal to the power dissipated in the resistor connected to the secondary (neglecting any power loss in the transformer or wires), so $I_p = (1/V_p) I_s V_s$ which is 1 A. The resistance that would draw 1 A when connected directly to 1000 V is 1000 Ω.

ELECTROMAGNETIC

WAVES I

Induction (Key 75) generalized a little bit, leads to the pre-diction that there are electromagnetic waves. In this Theme, we move quickly from the prediction to a description of one of the phenomena that comfirm it—light.

Key 77 Magnetoelectric induction in Maxwell's equations

OVERVIEW *A changing magnetic field produces a potential difference between two electrodes (Key 75). Maxwell also believed that a changing electric field would produce a magnetic field.*

Electromagnetic induction: A changing magnetic field induces an electric field, as Faraday, Henry, and many who repeated their experiments knew. Does a changing electric field do something magnetic? There were no experiments from the time of Faraday and Henry to the time of James Clerk Maxwell and Heinrich Hertz that suggested such a thing, but Maxwell's formulae summarizing the laws of electricity and magnetism did. Maxwell's equations make electricity and magnetism seem to be different phenomena without magnetoelectric induction but very similar phenomena with it. This is enough to compel a theoretical physicist to consider that electricity and magnetism may be very closely related. Maxwell recognized that if there is magnetoelectric induction then there must be electromagnetic radiation.

Magnetoelectric induction: Maxwell said that changing magnetic fields induce electric fields, and that changing electric fields induce magnetic fields. It does not require mathematical skill to see the point. If each changing field can induce the other, then the induction process may repeat, again and again. In an electromagnetic wave, changing electric fields make magnetic fields just ahead of themselves. Changing magnetic fields make electric fields ahead. Over and over, these processes make a wave that propagates ahead.
- A changing electric field is enough to start an electromagnetic wave. One way this may occur is if an electrically charged particle accelerates. Electrons can vibrate in atoms, so every atom is a potential source of electromagnetic radiation.

Key 78 Electromagnetic waves, their speed, and their spectrum

OVERVIEW *Radio waves, microwaves, radiant heat, light, X rays, and more: all are electromagnetic radiation, each with its own particular frequency.*

Wave speed: In the late 1800s, Maxwell's equations predicted that the speed of an electromagnetic wave would be 3×10^8 m/s. Only a few years after Maxwell's prediction. Heinrich Hertz observed phenomena in his laboratory consistent with the existence of electromagnetic radiation, but not consistent with any other explanation. Previous astronomical and new laboratory measurements of the speed of light, by Michelson and others, yielded the same speed. Hertz's radio waves and light waves are electromagnetic radiation.

Frequency: Any wave has a frequency, which is the rate at which the oscillating wave motion completes its cycles.
- Radio waves have frequencies of 10^6 to 10^8 Hz (cycles per second).
- Microwaves have frequencies from the radio range up to 10^{12} Hz.
- Infrared light, or radiant heat, has frequencies from 10^{12} to 10^{15} Hz.
- Visible light has frequencies close to 0.6×10^{15} Hz.
- Ultraviolet radiation has frequencies up to 10 times higher than visible light; X rays 10 to 1000 times higher.
- Gamma rays (γ-rays) have frequencies greater than 10^{17} Hz.
- Different kinds of waves were once thought to be different phenomena. All the types of radiation just mentioned differ only in their frequency; they are the same in every other aspect.

Wavelengths: Wavelengths can be computed from the frequencies and the speed of light given at the beginning of this Key. Wavelengths vary from a few meters to hundreds of meters for radio waves; tenths of a millimeter to a meter for microwaves; tenths of a micron to a millimeter for infrared waves; around 5×10^{-7} m for visible light waves; and still shorter for higher-frequency radiations.

Key 79 Transparent, opaque, and wavelength-selective materials

OVERVIEW *Electromagnetic waves pass through some materials (transparent) and die away in others (absorbers). Most objects do not emit visible light, but reflect some of the light that falls on them. Colored objects reflect some wavelengths of visible light but absorb others.*

Transparent materials: Visible light passes through transparent air or glass as if there were nothing there. Electrons in materials have natural frequencies at which they oscillate. Electromagnetic waves with frequencies much different from the natural frequencies pass by without making the electrons vibrate very much.

Opaque materials: Opaque materials contain electrons that oscillate naturally at the frequencies of visible light. When light enters these materials, their electrons respond resonantly and oscillate with large amplitudes. Frictional forces, though small, absorb a lot of power because of the large amplitudes. Waves with frequencies close to the natural frequencies of electrons lose energy rapidly; they are absorbed.

Metals: Light propagating in air toward a smooth metal surface is mostly reflected. Only the small part of the light energy that enters the metal is transformed to heat.

Wavelength and color: Wavelengths from about 400 nm to about 700 nm are visible to us. The longest, lowest frequency visible light waves appear red to most of us (color blindness is not rare, however). Shorter wavelengths (higher frequencies) appear in turn orange, yellow, green, blue, indigo, and violet.

Selective reflection and transmission: Roses are red, violets are blue (violet?), because they contain atoms and molecules that absorb all but red (or blue) wavelengths. The light not absorbed is reflected into many different directions by the rough surface. The rich colors we enjoy in our world all arise from this simple process that selects some of the colors in sunlight.

- Filters used to color theater lights, slides, and movies all use selective transmission to produce light rays with the required wavelengths picked to make the desired color appear on the screen or other objects they illuminate.

Blue sky and red sunset;
white clouds and green seas

OVERVIEW *Wavelength-dependent scattering of light by
air molecules or by small water droplets and wavelength-
selective absorption of light in water make some of the
familiar colors we observe in nature.*

Blue sky, red sunset: Visible light has wavelengths 1000 times longer
than the size of an air molecule. The air molecules have very little
effect on passing light, just as a single post in a harbor has little effect
on passing surface waves with wavelengths many times the post's
diameter. But longer red wavelengths are affected even less than
shorter blue wavelengths. In miles of atmosphere, more blue sunlight
is scattered to the side than red. This is why we see blue when we
look away from the sun at the sky.
- At sunset, the same scattering process makes the light reaching
you weak in the blue, relatively stronger in the red.

KEY ILLUSTRATION

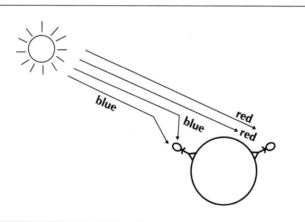

White clouds, green seas: The water droplets that make up a cloud are much larger in diameter than a visible light wavelength. Light rays hitting a drop scatter in all directions, longer and shorter wavelengths alike. Thus the sunlight scattered toward our eyes by a cloud is white.

- Beneath a smooth ocean surface only weak selective absorption by water affects the light in it. Meters of water pass green and blue but absorb red. Light that enters water and emerges again is blue-green.

OVERVIEW *How does light reflect from matter? One simple rule predicts how light rays move in reflection and in refraction (Key 83).*

Reflection at a smooth surface: Light, electromagnetic radiation, produces forces on the electrons in an atom. The electrons vibrate at the frequency of the light, in addition to the motion they would make in the dark. They emit light, producing rays in particular directions, determined by interference (Key 86). The angle of incidence equals the angle of reflection. A ray incident 25° away from the perpendicular reflects 25° away on the other side; the incident and reflected rays and the perpendicular all lie in one plane.

Fermat's principle and reflection: During the 1600s in France, Pierre Fermat discovered this rule: of all paths possible for light rays, they follow those that they traverse in the **least time**.
- The law of reflection stated above follows from this if you consider a light source and your eye in front of a mirror. Imagine the source to be where it appears to your sight, behind the mirror surface as far back as the real source is from the front. A straight line from this *virtual* source to your eye is the least-time path, and it crosses the mirror surface at the reflection point.

KEY ILLUSTRATION

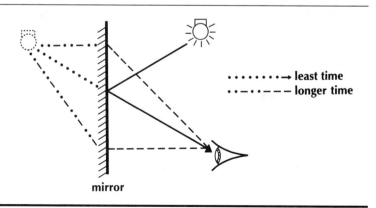

········→ least time
·—··—— longer time

mirror

OVERVIEW *The shape of a mirror's surface determines the apparent size and distance of an image seen in reflection.*

Plane mirrors: Rays from a point source of light in front of a mirror that reflect at different points on the surface all appear, after reflection, to come from a virtual point source behind the mirror. Using the rule of reflection given in Key 81, you can extend the reflected rays back in a scale drawing of the system and find how far behind the mirror the virtual source is. You find, again, that it is as far behind the mirror as the real source lies in front of the mirror.

- Using two real sources and finding their images by the method just described, you can show that the separation of their virtual images is the same as the separation of the real sources.
- A flat mirror produces a **virtual image**, through which no light rays actually pass, with **unit magnification**.

Curved mirrors: The same drawing of extensions of rays on a scaled diagram, always done according to the law of reflection stated in Key 81, can be used to demonstrate that a gently curved mirror produces magnification different from unity and an apparent distance from the mirror back to the virtual image different from the source to the mirror surface.

- A **convex** mirror (looking at it is like looking at a small piece of the surface of a large ball) produces an image smaller and closer than a plane mirror's.
- A **concave** mirror (looking at it is like looking into a cave or a bowl) produces a larger image at a greater distance than a plane mirror's.
- In some cases, concave mirrors produce **real images** through which the rays forming them actually pass. The curved reflector behind a flashlight bulb directs rays from the bulb away parallel to each other, producing an intense narrow beam. By reversing the direction of the rays in your mind, perhaps you can see why curved dish antennae are used to collect microwave radiation from distant transmitters and direct the rays to a small detector in front of the dish. Isaac Newton's reflecting telescope also used a concave mirror this way.

Key 83 Fermat's principle, the law of refraction, and total internal reflection

OVERVIEW *Light travels slower in water than in air and still slower in glass. Light traveling from a source in one medium to a detector in another medium changes direction at the interface when it follows a least-time path. Light approaching an interface between two media is partially transmitted through it and partially reflected. In a special case, **all** of the incident light is reflected.*

Refraction: Light traveling in air toward the flat surface of a piece of glass, in a direction 45° away from the perpendicular to the surface, **refracts** or changes direction when it enters the glass. Inside, it travels in a direction about 30° from the perpendicular. As in reflection, the incident ray, the refracted ray, and the perpendicular line all lie in a single plane. If the glass is a thin window pane with parallel surfaces, then the light returns to its original direction as it leaves through the second surface. To change the direction of a light ray in air by refraction, you need a piece of glass with plane surfaces that are not parallel.

Propagation speed: Light in air or vacuum moves at 3×10^8 m/s. In water, the speed is about 1/1.33 of the vacuum speed; in glass, about 1/1.5 of the vacuum speed. A straight line from a source in air to a detector in glass is not the fastest path available to a ray. A path that covers a greater distance in the air, but a smaller distance in the glass, takes less time because of the speed difference. The direction of a ray is closer to the perpendicular to the interface plane in a slower medium, farther from the perpendicular in a faster medium.

Critical angle: Light approaching an air-glass interface from the air side at grazing incidence, almost 90° from the perpendicular, travels in a direction about 43° from the perpendicular in the glass. This is the **critical angle** for glass. Consider light propagating along the same path in the opposite direction, and then consider increasing the angle away from the perpendicular in the glass. The angle in the air must also increase, but it cannot; any increase beyond 90° puts the emerging ray back in the glass. In this case, there is **no** direction in the air that satisfies the law of refraction, so there can be no rays in the air. All the light approaching the glass-air interface from the glass side at angles greater than 43° is reflected back into the glass.

KEY ILLUSTRATION

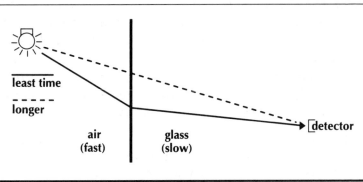

Applications:

- Total internal reflection is used in prisms in binoculars to fold a long optical path into a small package. Light approaches a glass-air interface from the glass side at an angle of 45° and is reflected without loss of intensity. A number of reflections bring an upright image of the viewed object to the user.

- The most important application of total internal reflection is in communications. Thin cylindrical fibers of very pure glass transmit light for miles without loss of intensity. Reflection from the smooth surfaces keeps all the light intensity in the fibers. Because light frequencies are so much higher than those used in electronics, much more information per second can pass over an optical data link than wires can transmit. City telephone companies that are running out of space in their conduits continue to meet increasing demands for their services by replacing their wires with optical fibers. Cable television can offer more channels and higher-resolution images by using optical fibers to carry their signals.

Key 84 · Prisms and lenses; dispersion

OVERVIEW *Light that enters glass from air through a plane interface and then leaves the glass **prism** through another plane interface that is not parallel to the first, travels in a direction after it has passed through the glass that is different from its original direction. Smooth, gently curved interfaces have a similar effect. The speed of light in water or in glass is not the same for all colors (wavelengths). The angle through which a prism deflects a light ray varies with the wavelength of the light. The consequences of this fact can be beautiful.*

Prisms: The following figure shows a light ray refracting at the plane surfaces of a glass prism according to the rule given in Key 83. The direction of the light ray is changed by the prism. The greater the angle between the two interfaces, the greater the deflection of the light ray from its original direction. Note that the light rays deflect away from the **vertex** where the two interfaces intersect, toward the **base** where they are most widely separated.

KEY ILLUSTRATION

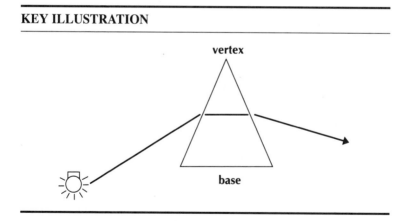

Lenses: A lens has at least one curved surface. The law of refraction is the same as for a plane surface (Key 83); the perpendicular to use in the law is the perpendicular to the curved surface at the point where the light ray crosses it. In a lens, nearly parallel interfaces near the center of the lens produce small deflections, and interfaces with larger angles between them away from the center produce larger deflections. Interfaces with shapes close to spherical deflect parallel rays so that they all cross at a single point beyond the lens (converging lens) or all move beyond the lens as if they had emanated from a single point before the lens (diverging lens); see the figure.

KEY ILLUSTRATION

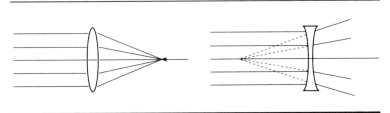

- Lenses can be used to form real or virtual images of objects, with magnifications that may differ from unity.
- A magnifying glass is a single lens that forms a virtual image with magnification greater than unity.
- Binoculars, microscopes, and telescopes use more than one lens to form images—and images of images—to do what we want them to do.

Dispersion: A glass prism deflects white light and spreads it out into a rainbow of colors. This would not happen if all colors (wavelengths) traveled at the same speed in the glass. But short violet wavelengths travel slower in glass than longer red wavelengths. The violets deflect more than reds, and the other colors are spread out in between. **Dispersion** means "spread out" or sorted according to wavelength.

Spectrographs: If a narrow beam of light enters a glass prism, and then the dispersed light emerging falls on a piece of photographic film, the film records the intensity as a function of wavelength. This **spectrum** is often used by scientists to characterize either the source emitting the light or the absorbing material between the source and the spectrograph.

Rainbows: Water disperses light as does glass. Spherical water drops change the direction of light rays that enter them by about 139°, 140° for violet light but only 138° for red light. In this process, the ray enters the drop with a change of direction, totally reflects internally once, and leaves with another change of direction. Under the right conditions, we see the refracted light as a rainbow with the colors spread out from violet lowest to red highest. In favorable conditions, a second bow appears higher with the colors reversed. This bow is produced by rays that reflect internally twice rather than just once in the spherical drops of water.

KEY ILLUSTRATION

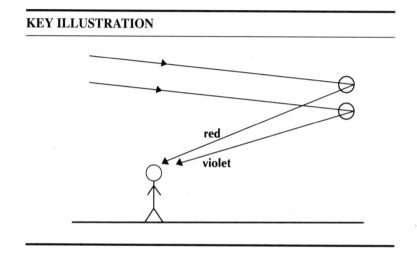

OVERVIEW *Sample questions of the type that might appear on homework assignments and tests are presented with answers.*

The speed of light:

- It takes six minutes for light to travel from Mars to the earth. How far away is Mars? $D = vt$. $D = 3 \times 10^8$ m/s $\times$ 6 min $\times$ 60 s/min or $D = 1.1 \times 10^{11}$ m.
- What is the wavelength of electromagnetic radiation that has a frequency of 10^9 Hz? The wavelength is $\lambda = c/f$ or 3×10^8 m/s/10^9 Hz, which is 0.3 m.
- What is the frequency of the blue light? Blue light has a wavelength of about 550 nm. Its frequency is $f = c/\lambda$ or 3×10^8 m/s/$(0.55 \times 10^{-6}$ m), which is about 5.5×10^{14} Hz.

Mirrors:

- You walk around toward a large flat mirror at a speed of 2 m/s. Your image in the mirror looks like it is walking toward you. How fast do you and your image seem to be coming together? The image appears to approach the mirror with the same speed as the person. Image and person approach each other at 2 m/s + 2 m/s = 4 m/s.
- You look into a mirror in a strange building at Six Flags over Mid-America. You see a person twice the height you know you are. What is the shape of the mirror? To show an image larger than the object, the mirror must be concave.

Refraction:

- When a wave crosses a plane boundary from a slower medium to a faster medium, does it bend toward or away from the line perpendicular to the boundary? When entering a faster medium, light bends away from the perpendicular.

Dispersion:

- What is the most important feature of a prism that is used to separate white light into colors? It is that each color (wavelength) has a different propagation speed in the prism.

Theme 14 ELECTROMAGNETIC WAVES II

Light waves were the subject of the previous Theme. In some situations, the wave behavior of light produces results different from those of simple ray description.

Huygens' principle, diffraction, and interference

OVERVIEW *Light wave crests move out like expanding spheres from a point source—like crests of water waves on the surface of a pond that expand circularly from the spot where a rock was dropped in. Waves interfere (Key 48) to produce complex patterns of large and small amplitudes when more than one source emits. Huygens explained reflection, refraction, and other ray phenomena with the simple idea that every point on a wave crest is a source of new, spherically spreading waves.*

Huygens' principle: Christian Huygens proposed a simple but successful idea in the late 1600s. In a wave that is propagating away from its source, what is the source of the new moving crests? Water waves suggest a simple answer. At any point, the water moves up and down as waves pass. If it did not move up and down in that region, then there would be no disturbance in nearby regions at later times. The source of later waves, therefore, is the present wave. A crest emits crests that spread outward in every direction at the wave speed.

- Applied to propagation of light, Huygens' principle successfully explains the spherical spreading of crests from a point source—the law of reflection from mirrors, for waves with crests that lie on planes (Key 81), and the law of refraction for plane waves crossing interfaces where the wave speed changes (Key 83). Huygens' principle also explains phenomena that require wave ideas, rather than simple wave ideas. Thomas Young's explanation of his two-slit interference pattern is the classic example.

- In its simple form given here, Huygens' principle cannot explain why a wave continues to propagate in one direction, rather than changing to two directions, forward and backward. Even with this gap, the pre-1700 rule is an impressive accomplishment.

Diffraction: Look closely at the shadow on a screen of a sharp edge illuminated by a distant light source. The shadow is not sharp, even though the edge is. If you think of the incident wave near the sharp edge as a source of new waves spreading out in all directions, then this is not too surprising; Huygens knew why this happened. Light of one color, best if it comes from a laser, falling on a thin slit produces diffraction: a spread of illumination wider than the slit on a screen behind it. Very close study reveals alternating minima and maxima within the decreasing intensity behind the slit edge, produced by interference between waves arriving at the screen from different points in the slit. As the slit is made narrower, the width of the illuminated region on the screen grows.

Interference: In a wave that is the superposition of waves from many sources, large amplitudes occur when and where the crests from single sources arrive at the same place at the same time. At other places and times, crests and troughs both arrive, tend to cancel, and produce smaller amplitudes. Constructive (all adding) and destructive (canceling) interference produce the intensity distributions observed in diffraction, in which sources with the same frequency interfere.

- Thomas Young demonstrated the wave nature of light just after 1800 by illuminating two pinholes with a single monochromatic (one color, one wavelength) source of light. On a screen beyond, there appeared a pool of illumination spread out by diffraction, not uniform but rather consisting of closely spaced lines of high and low intensity running perpendicular to the direction between the pinholes. At a point on the screen at which the difference in the distance to the two pinholes is one, or two, or any integer number of wavelengths, waves from both holes interfere constructively. At another point where the difference is an integer number of wavelengths plus another half, the waves interfere destructively. If the wavelength of the light is made longer, the pattern expands. Reducing the separation between the pinholes also expands the pattern. Using long, thin slits side-by-side produces the same pattern with more intensity on the screen.

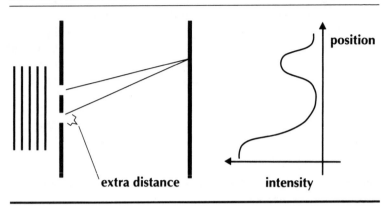

- A **diffraction grating** is an opaque sheet with many closely spaced slits. It produces very narrow peaks of high intensity at screen positions where waves from all the slits interfere constructively and low intensity in broad regions between the peaks. Grating spectrometers are instruments that spread light—sorted according to wavelength—out across a screen or detector.
- When light strikes two glass plates placed in contact, a thin film of oil floating on water on a street after a brief rainstorm, or the thin film surface of a soap bubble, an observer sees colors that change as the viewing angle changes. This is an interference effect. White light containing all wavelengths may reflect from either of two glass surfaces that are close, or from either the oil-air interface or the oil-water interface, separated only by the thickness of the oil film, or from either of the two soap-air interfaces. The paths from reflection points to the observer may differ by an integer number of wavelengths for one wavelength or an integer plus one-half for another. Colors for which constructive interference occurs appear to the viewer. With monochromatic illumination, the interference pattern produced by two glass surfaces, Newton's rings, is used to test the shapes of the surfaces.

OVERVIEW *Interference effects prove that light is waves. Polarization phenomena, which are easily observed, demonstrate that light waves are transverse, different from sound waves.*

Light: Longitudinal or transverse?: The electric and magnetic fields in propagating electromagnetic radiation are always perpendicular to the wave direction and perpendicular to each other as well. Light has transverse polarization (Key 47). The direction of the electric field vector is conventionally used to describe the polarization of light.

Polaroid: Imagine an AC electric field in a region of space filled with fine parallel wires, insulated from each other, extending up and down. An up-down field would drive currents in the wires and lose energy as those currents encounter frictional forces. A north-south field cannot make any current flow and so does not lose energy to friction. Polaroid is an atomic-scale array of such wires, molecules called polymers that are long chains of atoms along which electrons move freely. A sheet of Polaroid with its transmission axis oriented north-south, illuminated from the east by light with all possible polarization states, passes to the west only north-south polarized light, no up-down polarization.

- A second Polaroid sheet, farther west, with its transmission axis north-south, would pass all the light coming through the first sheet. Oriented up-down, it would stop all the light from the first sheet. At any other angle, it passes light with reduced intensity and with its polarization axis rotated to coincide with that of the second Polaroid.

- If you can obtain three sheets of Polaroid, you can produce a striking demonstration of the fact that Polaroid can change the orientation of polarized light. Begin with the first sheet oriented north-south, the third oriented up-down, and a little space between them. All light coming from the east should be stopped by the third sheet. Now insert the second sheet between the first and third sheets. If its transmission axis is aligned with that of either of the sheets, the system continues to stop all the light. But if its transmission axis is 45° from the others, it changes the orientation of the light that passes between the first and third sheets. The north-south light incident on it is partially absorbed, but some intensity passes with 45° polarization. At the third sheet, some of this light is absorbed, but again some passes with up-down polarization.

- Polaroid is used in some sunglasses. Glare and mirages (Key 83)—reflected or refracted light coming up from the land at a grazing angle on a bright, hot day—tend to be horizontally polarized. Glasses that pass only vertically polarized light reduce the intensity of glare relative to that of the light reflected from cars, signs, and other things we want to see.

OVERVIEW *In low-density gases with electric currents flowing through them, atoms absorb energy from the current and sometimes re-emit the energy as electromagnetic radiation. Gases without flowing currents absorb some of the light passing through them. Both processes occur at a few wavelengths characteristic of the atoms in the gas, not at other wavelengths. If light of its characteristic frequency is present, an excited atom is more likely to emit, and the emitted light will interfere constructively with the already-present light. These phenomena have many applications and allow scientists to study the dynamics of the electrons in atoms.*

Atomic emission and the electronic states of atoms: low-density, monatomic (one atom per molecule) gas, neon, in this example, can be excited by passing a current through it and will emit orange light. A prism spectrograph (Key 84) or grating spectrometer (Key 86) can be used to determine the emission spectrum of neon, the wavelengths at which neon emits light. There are only a few in the visible range, separated by broad wavelength regions in which no light is emitted. This suggests an outline of the dynamics of electrons in atoms. They act as if they can be in states with particular energies and change state by emission of radiation (Key 94). There are not electronic states of the atom at every energy, but only a few possible states separated in energy by broad ranges in which no states exist.

Absorption: A low-density gas that is not excited by an electric current, not heated, will absorb light at a few characteristic wavelengths, not at others.

- An atom's characteristic wavelengths for absorption, its absorption spectrum, are the same as those for emission.
- An early application of this phenomenon in the study of the nature of the sun was made by J.D. Fraunhofer, a Bavarian who observed that white light from the sun had some wavelengths missing. The pattern of missing wavelengths showed that the cooler outer atmosphere of the sun contains some atoms already identified on the earth and one never seen here—helium. It was found on the earth after its discovery in the sun.

Incoherent light: Light emitted spontaneously by a low-density excited gas—even if there is only one frequency, one wavelength present—has less than the maximum intensity possible because of the destructive interference. If you could arrange for all the atoms to emit crests at times so that the interference was constructive, you could produce a higher intensity than what randomly emitting atoms produce. Ordinary light, from electric lamps or the sun, has this random character and is called **incoherent**.

- Light with all waves interfering constructively is called **coherent**.

Coherent light and stimulated emission: To be coherent, light must have intensity at only one frequency or wavelength, and the crests from different sources must line up perfectly. Spontaneous emission from atoms does not produce this alignment. Stimulated emission, in which radiation already present produces forces on electrons in an atom and gets them moving in step, leads to emitted light in phase (crests line up) with the background light. Introducing low-intensity light at the characteristic frequency into a gas of excited atoms can make a large number of them radiate so that the electromagnetic fields in their waves add without canceling to produce a very high-intensity wave. A. Einstein predicted stimulated emission in 1917.

Lasers: **L**ight **A**mplification by **S**timulated **E**mission of **R**adiation: the name of the machine describes the process it facilitates. In a laser, a low-density gas or other material is excited by an electric current, by absorption of light, by a chemical reaction that releases energy, or by more exotic means. Spontaneous emission produces a low level of light at the laser frequency. It stimulates further coherent emission. This does not happen in an ordinary gas-discharge lamp because the intensity never becomes high enough to make stimulated emission more likely than spontaneous incoherent emission. In a laser, mirrors at the ends of the emitting material reflect light back into the material; they make the intensity higher in the material than it would be if light passed through only once. One mirror is only partially reflecting, so some light escapes. Under these conditions, laser action—stimulated emission—occurs. The light so produced is very intense, very monochromatic (the intensity occurs in a narrow range of frequency or wavelength), and forms a beam that increases its diameter slowly as it propagates away, rather than the rays spreading out in all directions that an incoherent source produces.

Key 89 Black-body radiation

OVERVIEW *Solids heated to high temperature, like the filament in an incandescent bulb, emit a smooth spectrum (that is, a graph of intensity versus wavelength is a smooth curve). The temperature dependences of the power emitted and of the spectrum reveal information about the dynamics of electrons in condensed matter.*

Incandescence: Hot solid tungsten glows red, orange, or white as temperature increases. A low-density gas of tungsten atoms emits radiation at just a few wavelengths. The solid clearly behaves differently from the gas. Spectroscopic measurements have shown that the frequency ($f = c/\lambda$, Keys 46 and 77) at which the radiation from an incandescent source is most intense, the peak frequency, increases as the source's temperature increases: $f_{peak} \propto T$. The total power radiated at all frequencies increases with temperature: $P \propto T^4$.

A black body: The more radiation a body absorbs, the more it must emit if it is in thermal equilibrium with its surroundings (Keys 36 and 39). A perfect black body is one that absorbs all radiation falling on it; at any temperature it both absorbs and emits more radiation than a shiny object. Statistical mechanics, developed in the nineteenth century, predicted some, but not all, of the features of the black-body spectrum correctly.

In the Key Illustration the horizontal axis is frequency f in units of 10^{14} Hz, and the vertical axis is the frequency distribution of the intensity of black-body radiations $R(f)$ in units of 10^{-9} Wm^{-2}Hz^{-1}. From the lowest to the highest, the curves are for temperatures of 1000 K, 1500 K, and 2000 K.

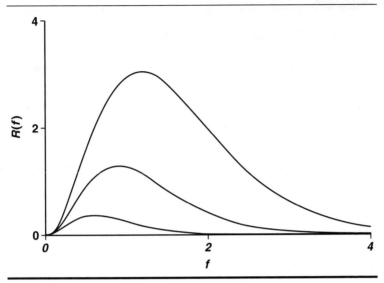

OVERVIEW *Sample questions of the type that might appear on homework assignments and tests are presented with answers.*

Diffraction:
- Consider the two-slit diffraction pattern of the figure in Key 86. Assume it is produced by red light. How would it change if blue light were used instead? How would it change if red light were used with the separation between the slits reduced to half its original distance? Blue light has a shorter wavelength than red light, and so the intensity maxima on the screen would be closer together with blue light. Reducing the separation between the slits while holding the wavelength constant requires a greater angle between intensity maxima for the "extra distance" in the figure to be one wavelength. Reducing the slit separation increases the separation between intensity maxima on the screen.
- Coherent, monochromatic (one color) light from two closely spaced slits produces a diffraction pattern. Does light from two stars that appear to be close together do the same? No. Even if you use a filter to allow only one wavelength through, there is no two-source diffraction pattern because the light from one star is not coherent (crests lining up with crests) with light from another star.

Polarization:
- Consider three sheets of Polaroid, lined up with their transmitting axes parallel, in a beam of unpolarized light. The intensity coming through is now I_0. Now rotate them so that the first sheet is vertical, the second is at 45°, and the third is horizontal. What is the power coming through? If the polaroid at 45° is removed, what power comes through? With three sheets and 45° angles, the power coming through is less than I_0 but greater than zero. With two sheets at 90°, no power comes through.

Classical physics—mechanics, electromagnetism, thermodynamics, and statistical mechanics—explained many of the phenomena observed in ever-more-precise experiments throughout the nineteenth century. In just a few cases, it failed to explain phenomena or produced obviously wrong results. Attempts to "fix up" classical physics failed. Physicists who refused to ignore the problems created the spectacularly successful quantum physics, which today explains phenomena that occur with length scales 10^{-10} times smaller than those for which it was invented.

Key 91 The black-body spectrum and Planck's constant

OVERVIEW *A black body radiates energy at all frequencies but mostly near a peak frequency proportional to its temperature. It radiates much less energy at very low and very high frequencies. Classical physics wrongly predicts ever-increasing energy radiated at increasing frequencies, and worse, a total radiated power that is infinite at any temperature. Planck "fixed up" classical physics so artificially that he was embarrassed to publish his result.*

Observed and calculated black-body spectra: A graph of electromagnetic power radiated at each frequency by a hot black body versus frequency is a smooth curve with a single broad peak (Key 89). The spectrum can be calculated using mechanics, thermodynamics, elementary statistics, and a model of the body in which randomly oscillating electrons emit and absorb radiation. The result is a smooth curve that rises higher and higher as frequency increases, with a total emitted power infinite at any temperature above absolute zero. The world does not behave this way.

Planck's better calculation: In 1900, the German physicist Max Planck published a calculation that agreed in every detail with the observed black-body spectrum. He assumed that matter and electromagnetic radiation at frequency f cannot exchange arbitrarily small amounts of energy, but that no matter what the temperature and no matter what the radiation intensity, energy is always exchanged in units of fixed size—**quanta**. Large amounts of energy can be exchanged by emission or absorption of a large amount of quanta, but there can be no energy exchange smaller than one quantum of energy. The energy E in a quantum depends on the frequency f:

$$E = hf$$

where h is **Planck's constant**,

$$h = 6.63 \times 10^{-34} \text{ Js.}$$

- The value of h is so small that no experiment known to Planck could directly reveal the discrete, as opposed to continuous, set of energies that could be exchanged under the assumption.

Key 92 Einstein's theory of photoelectric effect

OVERVIEW *Light shining on a clean metal surface in a vacuum (cathode) can cause a current to flow through the vacuum to another dark electrode (anode). At low intensity, the current varies with frequency and intensity in a way different from the predictions of classical physics.*

Photoelectric experiments: Low-intensity light of any frequency should, after enough time, deposit sufficient energy in the illuminated cathode to break an electron loose. But the current, if observed at all, always began as soon as the illumination was turned on, no matter how low its intensity. For low-frequency light there was no current; the light had to have a frequency greater than a minimum value to produce current even if its intensity was very great. And the maximum kinetic energy that an electron could acquire measured by manipulating the voltage between the anode and the cathode and finding a critical voltage at which the current dropped to zero, depended on the frequency of the light but not on its intensity. Classically, one would expect the critical voltage to depend on intensity, the light energy crossing a unit area per second. One would expect delayed current at very low intensity, but always some current eventually, regardless of the frequency.

Einstein's theory: Investigating how Planck's quantum hypothesis (Key 91) would alter classical expectations about the photoelectric effect, Albert Einstein found that it explained all the observations. The cathode absorbs energy from the light one quantum at a time. If one quantum of energy is not enough to break an electron free, it spreads and turns to heat quickly, before another quantum is absorbed at the same place. Thus the light frequency f must be high enough that the energy hf of a quantum is enough to remove the electron from the cathode. The maximum kinetic energy an electron could acquire would be the quantum energy less that needed to escape from the cathode and so would increase with increasing f regardless of the intensity. Low-intensity light would deliver fewer quanta per second than high-intensity light, but assuming that quanta arrived at random times, one could arrive just after the light was turned on no matter how low the intensity.

Key 93 Wave-particle duality and complementarity

OVERVIEW *Light behaves as a wave, as interference experiments prove. But in the photoelectric and black-body radiation phenomena, light behaves more like a shower of tiny particles, each with one quantum of energy. Particles like electrons behave like waves sometimes (Key 95). Quantum objects behave like both.*

Wave-particle duality: Waves behave like particles when they interact with a hot body, and as we shall soon see, particles behave like waves. The view of this situation that developed in the quantum revolution is that all quantum objects exhibit both behaviors, one or the other depending on the details of the experiment. This is called **duality**. Clever physicists have arranged experiments in which both behaviors occur.

- Young's two-slit diffraction with its interference fringes can be measured today with image intensifiers and television cameras so sensitive that they record the arrival of a single quantum. With very low intensity illumination, a display showing where each detected quantum arrived is dark as the experiment begins. Bright spots slowly appear on the display at random times and in apparently random positions at the beginning. As more and more quanta are recorded, a pattern emerges on the display that was not evident at the start; most of the quanta arrive at the location of Young's bright fringes, very few at the dark fringes. Each quantum particle is affected by the two slits as a continuous wave would be.

Complementarity: Complementarity means a physicist cannot correctly consider only particle behavior or only wave behavior. In the right experiment, which must involve high frequencies because Planck's constant h is so small, light waves are both waves and particles. The quanta are called **photons** to emphasize the particle nature of light. Niels Bohr, the Danish physicist who formulated the first quantum model of atomic structure (Key 94), anticipated this when he first expressed complementarity. He insisted that seemingly opposite and contradictory phenomena are aspects of a deeper unity of behavior, long before his peers were prepared to believe such a thing of wave and particle behavior.

Atomic structure: Rutherford and Bohr

OVERVIEW *An experiment done in 1909 revealed the concentration of atomic mass and positive charge into a small central nucleus. Spectroscopists had described many regularities in atomic spectra by 1909 that revealed something of their electronic structure. Knowledge of electrons, nuclei, spectra, and the new quantum ideas challenged physicists to conceive a model for the atom.*

Scattering experiments: British physicist Ernest Rutherford was able in 1909 to direct a beam of helium nuclei toward a thin free-standing film of solid gold. He measured how many beam particles were deflected or scattered by the gold in different directions. If the mass and positive charge in atoms occupied the same volume as the electrons, the scattering would have been concentrated in a small cone centered on the forward (unscattered) direction. Instead, a few scattered particles were detected even at angles near $180°$, as if there had been a head-on collision with something much more massive than the beam particle.

- The bouncing of the beam particles from full atomic masses implied that the mass of a gold atom was concentrated in a very small volume. Nuclei have radii around 10^{-15} m, 10^{-5} times smaller than atomic radii.

Spectra and Bohr's atomic model: Spectroscopists J.J. Balmer, J. Rydberg, W. Ritz , and others found simple formulae describing some of the characteristic frequencies of light from simple atoms like hydrogen. Danish physicist Niels Bohr applied quantum ideas to the newly found nuclear picture of atomic structure. He showed that if electrons were in Kepler orbits around the hydrogen nucleus, restricted to angular momenta that are integer multiples of $h/2\pi$, the differences between the calculated energies of these orbits corresponded to the energies hf of photons emitted and absorbed.

- Bohr used this model to explain the energies of characteristic X rays emitted by atoms heavier than hydrogen. He had made a good start toward a quantum theory of atomic structure and dynamics.

de Broglie's electron waves and electron diffraction

OVERVIEW *Why should electrons in atoms be confined to the orbits Bohr described? A French graduate student proposed a then-wild idea in his Ph.D. thesis in 1924. Electrons behave like waves in atoms, and they form interference patterns when they reflect from the evenly spaced planes in a crystalline solid. Baseballs, bullets, and automobiles do not exhibit any detectable wave behavior.*

If waves are particles, are particles waves? Louis de Broglie explored the consequences of a positive answer to this question in his Ph.D. thesis. He argued that the states of the electron in hydrogen should be stationary waves, like the standing waves on a stringed musical instrument. The circumference of a Bohr orbit should be an integer number of wavelengths. He found that, if he assumed electrons had a wavelength given by

$$\lambda = \frac{h}{p}$$

where h is Planck's constant and p is the momentum of the electron, then his standing-wave condition produced the same orbits that Bohr's model did.

Are particles waves? de Broglie's success strongly implied that electrons should exhibit wave behavior in the right experiment, and he told the experimenters how to calculate the wavelength. It is typically very small, less than 1 nm for an electron with a kinetic energy around 10 eV. Higher-energy electrons have even shorter wavelengths. With wavelengths less than 1/500 those of the light, electrons do not clearly exhibit wave behavior in, for example, a Young's interference apparatus with dimensions that work for light. There were no data to test de Broglie's idea when he published it.

X-ray diffraction: It was known in the 1920s that X rays, photons with wavelengths shorter than 1 nm, produce complex interference patterns when they reflect from planes of atoms in crystals. Like a diffraction grating (Key 86), a crystal with thousands of planes produces intense diffraction peaks in a few directions in which all the reflected waves interfere constructively and low intensity in all other directions.

Electrons: With de Broglie's relation for the wavelength of a quantum electron, experimenters could produce a beam of electrons and direct it at a clean crystal surface in a vacuum. C.J. Davisson and L. H. Germer observed an electron diffraction pattern in such an experiment in the United States in 1927. The angles between the intense peaks were those expected, knowing the crystal structure and de Broglie's wavelength.

Modern experiments: Just as photons exhibit both wave and particle behavior in a modern version of Young's experiment (Key 93), so do electrons. An electric field called an electrostatic biprism functions as would two closely spaced slits, and the same sensitive imaging equipment described in Key 93 detects each arriving electron. Each electron interacts with both sides of the biprism; you cannot say that it passed through one side or the other. Yet each excites the detector at a single point.

Key 96 The uncertainty principle

OVERVIEW *If particles behave like waves, and they do, then how can one say where a particle is? You cannot specify both position and momentum simultaneously, nor energy and time of an event simultaneously, when dealing with quantum objects. This question never arises in the macroscopic world in which the smallness of Planck's constant makes quantum effects unobservable.*

Position uncertainty: You can say where the center of a billiard ball is. But you cannot say where an electron or a photon was when it was affected by two slits in Young's interference experiment. If you cover one slit and not the other, you know where electrons or photons you detect came from, but there is no longer an interference pattern. If you see an interference pattern, you cannot say an electron or a photon went through one of the two slits. The uncertainty in the location of any quantum particle is related to its wavelength; you can say where it is to within some number of those wavelengths.

Heisenberg's principle: In 1927, the year electron diffraction was first observed, the German quantum physicist Werner Heisenberg used de Broglie's relation and the impossibility of specifying one point as the location of a wave to show that the complete knowledge of a particle's present state and future motion, possible in classical mechanics, is never possible for quantum objects. Using the uppercase Greek letter, Δ, to represent a small range of values of the quantity who symbol follows it, Heisenberg showed that

$$\Delta x \Delta p \geq \frac{h}{2\pi} \approx 10^{-14} \text{ Js}$$

where h is Planck's constant. An electron in a hydrogen atom in its lowest-energy state, its ground state, is somewhere within a volume about 0.1 nm in diameter centered on the nucleus. In the frame of reference in which the nucleus is at rest, so is the electron; its momentum is near zero. But Heisenberg's uncertainty principle states that the momentum is not exactly zero. It fluctuates and has magnitudes as great as 10^{-24} kg $\cdot$ ms^{-1}. This corresponds to speeds as great as 10^6 ms^{-1} and kinetic energies as great as a few eV. These numbers are consistent with those you would obtain from the lowest-energy Bohr orbit.

- Baseballs, automobiles, and people can be located with precision that varies, but may be as good as 0.1 mm = 10^{-4} m.
- Optical microscopes locate smaller objects with a precision of about 10^{-6} m.
- The momenta of moving macroscopic objects are on the order of 1 kg • ms^{-1}.
- The quantum uncertainty in the momentum of a macroscopic object is about 10^{-28} kg • ms^{-1}, much smaller than a typical momentum and also much smaller than the precision with which experiments could determine the momenta of macroscopic objects.
- We see no quantum behavior of baseballs or people—no uncertainty in position or speed, no diffraction, because Planck's constant h is so small.

Energy-time uncertainty. Consider a massive quantum particle acted on by a force F that changes its momentum while it moves through some distance

$$\Delta x \Delta p \geq \frac{h}{2\pi} \qquad\qquad \Delta x (F\Delta t) \geq \frac{h}{2\pi}$$

The uncertainty in the momentum change is expressed as the product of the force and the uncertainty in the time the process took. Rearranging,

$$(F\Delta x)\Delta t \geq \frac{h}{2\pi} \qquad\qquad \Delta E\, \Delta t \geq \frac{h}{2\pi}$$

The uncertainty in the energy change multiplied by the uncertainty in the time for the process must, just like the product of position and momentum uncertainties, be greater than a minimum value given in terms of Planck's constant. For massless particles, photons, Planck's and de Broglie's relations can be used to compute momentum and energy from the frequency, with the result $E_{photon} = cP_{photon}$, where c is the speed of light. These uncertainty principles apply to photons and electrons alike (i.e., to all quantum wave/particles).

- If an excited atom emits a visible photon in a very short time interval, we know when the photon was "born" with an uncertainty equal to the duration of the emission process Δt. This time is typically 10^{-9} s. Using the energy-time uncertainty principle, this implies that photons emitted by a collection of identical atoms will not have exactly the same energy, but will have a range of energies $\Delta E \geq 10^{-25}$ J. The uncertainties in frequency and wavelength are 1.6×10^8 Hz and 1.6×10^{-4} nm, about 3×10^{-7} times the 550 nm wavelength.

Key 97 Schrödinger's equation, quantum mechanics, and the correspondence principle

OVERVIEW *Physicists believe they understand a wave when their theories produce a wave equation, but there was no such equation for de Broglie's waves when he proposed them. Laws of quantum behavior seem to contradict classical physics; must classical physics then be discarded?*

Wave equations: From Newton's laws one can derive an equation that describes the relationship between the rate of change of pressure at one point in a gas and the change in pressure with position at one instant in time. These rates are related in such a way that the mathematical solution to the equation describes a propagating pressure wave. This "physics pedigree" of sound waves makes scientists feel they understand sound. The original quantum physicists felt no such confidence about the emerging picture of quantum waves.

The quantum wave equation: Erwin Schrödinger, and Austrian physicist, discovered a wave equation whose solutions can be interpreted as de Broglie's waves. What waves is a wave function, Ψ, a function of position and time. Its square, $|\Psi|^2$ can be interpreted as the probability that an electron is at a certain position at a given time. The new quantum theory seemed to replace the determinism of classical physics with a system that could only predict probabilities. Albert Einstein was among many who objected to this situation, criticizing quantum theory as incomplete. Others believed that the universe is intrinsically unpredictable at the quantum scale of lengths and times. Quantum theory has survived the criticism and has proved accurate in tests made with length scales 10^{-10} times smaller than those for which it was formulated.

The correspondence principle: Niels Bohr first stated this rule, good for all scientific revolutions. Although quantum theory and classical physics seem contradictory, classical physics is accurate in the macroscopic world. In situations in which the old theory works, the new theory must produce the same predictions as the old. In human-scale situations, Schrödinger's equation and Newton's laws do make identical predictions.

163

OVERVIEW *Sample questions of the type that might appear on homework assignments and tests are presented with answers.*

Planck's relation:
- If you exchange an old photon for a new one with a wavelength four times as large, how does the new energy compare with the old? The new frequency is one fourth of the old, and so the new energy is one fourth of the old.
- What is the kinetic energy acquired by an electron initially at rest when it absorbs one blue photon? A blue photon has a wavelength of about 550 nm and so a frequency of about 5.5×10^{14} Hz (Key 85). The energy transferred is $E = hf$, where $h = 6.6 \times 10^{-34}$ Js, so $E = 3.6 \times 10^{-19}$ J or 2.3 eV (Key 59).

The photoelectric effect:
- Consider the photoelectric effect, illuminated with blue ($\lambda = 550$ nm) light. Imagine replacing it with a green ($\lambda = 450$ nm) light that emits the same number of photons per second. How does the number of electrons knocked out of the metal in one second change? How does the maximum kinetic energy of a photoemitted electron change? The number of electrons per second does not change; it still takes one photon per emitted electron, even if the photon's energy increases. The maximum kinetic energy of an emitted electron is greater when more energetic green photons are used.

de Broglie's relation:
- An electron has a momentum of 6.6×10^{-30} kg • m/s. What is its wavelength? de Broglie's relation is $\lambda = h/p$ or $\lambda = 6.6 \times 10^{-34}$ Js/6.6×10^{-30} kg • m/s, which is 10^{-4} m.

The uncertainty principle:
- Looking at a hair in a microscope, you can see that it moves no more than 10^{-6} m as you watch. The mass of the hair is $m = 10^{-8}$ kg. How big must its speed be, according to the uncertainty principle of Heisenberg? Momentum p is mv, where v is speed. $v = p/c \geq h/2\pi m\Delta x$ which is 10^{-34} Js/(10^{-8} kg $\times 10^{-6}$ m) or 10^{-20} m/s.

Theme 16 RADIOACTIVTY, NUCLEAR PHYSICS, AND PARTICLE PHYSICS

This Theme describes energy exchanges between atomic nuclei and electromagnetic fields, between tightly bound electrons in heavy atoms and electromagnetic fields, and between fragments of nuclei. The energies are higher than those of atom-visible light interactions by a thousand times or more. The nuclear fragments have been divided and divided, again and again, in the ongoing search to discover the most fundamental particles of which all others are composed.

Key 99 X rays and radioactivity

OVERVIEW *Rays that could darken photographic film (after development) like light but could pass through solid matter were discovered in the 1890s. Ionizing radiation, these rays are sufficiently energetic to dissociate electrons from positive ions in matter they traverse.*

X rays: The German physicist Wilhelm Roentgen discovered that a beam of electrons in a vacuum would, if stopped by a glass or metal wall, emit rays that penetrate through solid matter, expose photographic film as ordinary light does, and cause some materials to emit light. Medical applications for visualizing the interior of a patient's body followed almost immediately. An understanding of the nature of X rays took longer. They are high-energy photons with energies greater than about 10^3 eV, rather than the 2 to 3 eV of visible light.

Other new rays: Antoine Henri Becquerel, a French physicist, inquired shortly after Roentgen's discovery whether any elements emit X rays spontaneously, using photographic film. Most elements produced no effect, but uranium, thorium, and actinium did emit some kind of rays, which proved to be different from X rays.

Alpha, beta, and gamma rays: The rays emitted by heavy elements behave in three characteristic ways when they pass through a magnetic field perpendicular to their velocities. Alpha rays, positively charged and massive, change direction slightly. Beta rays, negatively charged and less massive, bend the other way and farther. Gamma rays are not deflected. Alphas are helium nuclei—two protons and two neutrons bound together. Betas are high-energy electrons, and gammas are photons with energies even higher than those of X rays. Their ability to penetrate matter varies. Five pages of this book is enough material to stop most alphas. Betas would go right through them but not a sheet of metal. Gammas require many centimeters of a heavy element like lead to stop them.

Radioactivity: Elements that spontaneously emit these new rays are called radioactive. The rays are too energetic, many keV to many MeV, to be produced by any atomic electrons; they must come from the nuclei. The nuclear reactions they reveal change nuclei of one element to those of another. The product nuclei may form in excited states and later emit gammas.

OVERVIEW *Nuclear radioactivity revealed the structure
and some of the dynamics of the nuclei of atoms.*

Nucleons: Nuclei contain protons, positively charged particles with mass
about 2000 times the electron mass and an equal and opposite charge
to the electron's. In 1932, James Chadwick demonstrated that rays
produced in his apparatus had a mass nearly the same as a proton's and
no charge. These are the neutrons that, along with protons, make up a
nucleus. The helium nucleus, containing two of each, is very stable.
- Alpha emission from heavier nuclei proves that nuclei can break
  into two or more lighter nuclei.
- In beta emission a neutron in a nucleus transforms into an electron
  and a proton, with the release of some energy.

Nuclear forces: A nucleus containing only neutral and positively charged
particles is electrically unstable. Many nuclei never fly apart, so there
must be some nonelectrical force holding the nucleons close together.
Gravity is much too weak (Key 54). The binding force has been named
the strong force. Decay, rare even in many unstable nuclei, must be
caused by something not as strong; it has been named the weak force.

Nuclear and particle physics: More and more nuclear reactions, and
more and more "fundamental" particles, too, were discovered in
experiments like Chadwick's, in which nuclear rays from one element
were directed at another element, and a particle detector recorded the
presence or absence of nuclear radiation. Later experiments used electromagnetic machines to accelerate charged particles to very high
energies (10^{12} eV in each of two colliding beams at Fermilab, 7×10^{12}
eV in each coming soon at CERN). They may be directed at station-
ary targets or made to collide with similar particle beams traveling in
the opposite direction. Out of these high-energy collisions come many
new particles. Recently, the curiously named theory of the strong
force, quantum chromodynamics, along with the now well-established
electroweak theory, together called the **standard model,** has
explained much of particle physics in terms of a relatively small number of elementary particles: quarks and gluons, of which neutrons and
protons are made, and the electron and a few of its relatives.

OVERVIEW *Unstable nuclei decay by radioactive processes. The decays occur randomly in time. One unstable nucleus is unpredictable, but a large number can be characterized by statistical methods. One of the most important applications of nuclear decay has been measurement of the ages of objects studied by geologists, archeologists, anthropologists, space scientists, and many others.*

Exponential decay and half-life:

- In a large number of unstable nuclei, a well-defined fraction of them will decay in the next second (or hour, year, thousand years—whatever time scale is appropriate). At the end of the second, there are fewer unstable nuclei, but otherwise everything is the same as before. In the following second, then, the same fraction of the now-smaller number will decay. Second after second, the same fraction of the remaining supply of unstable nuclei decays.

- The half-life of an unstable isotope is the time it takes for half of an original large number of nuclei to decay. Half-lives vary from less than 10^{-6} s to more than 10^9 years. The fraction of the starting number of nuclei left after one, or two, or 19 half-lives is the same for every unstable isotope, independent of its identity or the magnitude of its half-life. After one half-life, $2^{-1} = \frac{1}{2}$ of the starting number remain. After two half lives, $2^{-2} = \frac{1}{4}$ is the fraction remaining. After three half-lives, $2^{-3} = \frac{1}{8}$ remains. After N half-lives, $2^{-N} = \frac{1}{2^N}$ remains.

Radiocarbon dating: In addition to stable $^{12}_{6}C$ in CO_2 molecules in the earth's atmosphere, about one in every 10^{12} carbons is radioactive $^{14}_{6}C$. These decay by beta emission with a half-life of 5730 years:

$$^{14}_{6}C \rightarrow \, ^{14}_{7}N + \, ^{0}_{-1}e$$

Their concentration does not halve every 6000 years or so because cosmic rays—radiation from outer space—always generate more. The generating reaction, driven by cosmic-ray neutrons, is

$$^{1}_{0}n + \, ^{14}_{7}N \rightarrow \, ^{14}_{6}C + \, ^{1}_{1}H$$

The concentration of one $^{14}_{6}C$ in every 10^{12} carbons results from a balance between the decay and generation processes just listed.

- Living plants and animals constantly exchange carbon atoms with the atmosphere and so carry the same fraction of the radioactive $^{14}_{6}C$. One cubic centimeter of living matter contains enough radioactive carbon to produce a few beta rays per minute. At death, the exchange of carbon between organic matter and the atmosphere ceases. The fraction of radioactive carbon in organic matter decreases exponentially from the moment of death. By separating a standard mass of carbon from dead organic matter and measuring the rate at which it produces betas, you can determine how long ago the organism died.

- Radiocarbon dating is useful for dates as long as 50,000 years ago. In older matter, there are so few radioactive carbons left that measurement of the very few betas produced requires inconveniently long times.

Accelerator mass spectrometry: A sample of carbon that produces only one beta every few weeks may still have many $^{14}_{6}C$'s in it. They are hard to detect by their decay because their half-life is so long. In a more sensitive dating technique, the carbon is vaporized, ionized, accelerated by electric fields, and deflected by magnetic fields. The heavier $^{14}_{6}C$ bends less in the magnetic field than the lighter stable carbon. Detectors count each arriving ion and measure the ratio of the concentrations of the carbon isotopes using all the radioactive carbons, not just the few that decay during the measurement.

Other dating techniques: In geology and space science, isotopes of other elements such as uranium can be used to date samples. The dates obtained are always times during which the sample was isolated from other sources of the dating element and during which the isotopic composition changed only by radioactive decay. The users of these techniques must have some means of knowing the starting isotopic composition. Dates as great as a few billion years have been determined.

OVERVIEW *Fission is a nuclear reaction in which a heavy nucleus fragments into two intermediate-mass nuclei and a few odd particles, releasing 10^8 times the energy per atom released in a typical chemical reaction.*

The fission reaction: Nuclear forces, the strong and the weak forces, are short ranged. In very large nuclei, protons near the surface are repelled electrically almost as strongly as they are attracted by the strong force. Excited so that it vibrates, changing shape, the nucleus may fragment. The first fission reaction observed

$$\, _{0}^{1}n + \, _{92}^{235}U \rightarrow \, _{36}^{92}Kr + \, _{56}^{142}Ba + 3(\, _{0}^{1}n)$$

was detected by German scientists Otto Hahn and Fritz Strassman in 1939. They bombarded uranium with neutrons hoping to synthesize heavier nuclei but found unexpected barium. Lise Meitner and Otto Frisch, German scientists working in Sweden, first proposed the fission reaction as the explanation.

Energy production: A chemical reaction between two atoms may release an energy like 2 eV. The fission reaction described above releases 2×10^8 eV, mostly in the form of kinetic energy of the products. In agreement with an equation found by Albert Einstein in his special theory of relativity (Theme 17),

$$E = mc^2$$

where c is the speed of light, and the source of the energy is a small mass difference between the reactants and the products. A small loss of mass produces a huge energy.

Chain reaction: The fission reaction described above requires one neutron and releases three product neutrons. This fact suggested the possibility that a self-sustaining process might occur in uranium, in which fission product neutrons cause further fission reactions until all the $_{92}^{235}U$ has reacted. Even before it was demonstrated, this process was given the name **chain reaction**. Enrico Fermi, an Italian physicist working in Chicago, and his coworkers produced the first controlled chain reaction in 1942. To do so, they had to assemble a **critical mass** of uranium, enough to make its surface-to-volume ratio sufficiently small so that product neutrons would not just escape without further reaction. They also had to include a **moderator** in their structure to slow the product neutrons without absorbing them; slow neutrons are more likely to be absorbed by uranium nuclei than fast ones. They also had to include movable **control rods** of a neutron-absorbing material, so they could stop the reaction at will.

- In modern electric-power-generating reactors, the energy released heats and evaporates water to produce steam that drives turbines and electric generators. The engineering done to avoid release of radioactive material makes these reactors complicated and expensive.

- In a fission bomb, there are no control rods and no moderator. Pieces of less than the critical mass are assembled quickly to make a larger-than-critical mass in which the reaction races unchecked. The rapid release of a large amount of energy produces a tremendously powerful explosion.

OVERVIEW *This Key describes the small mass differences between sets of isolated nucleons and assembled nuclei, which supply the energy in nuclear reactions.*

Measuring mass: Key 101 includes a description of mass spectrometry, which is the separation of isotopes according to their masses. Done carefully, this technique has produced very accurate and precise data on all the elements. The total mass of a nucleus increases with increasing atomic mass number, but not exactly linearly. The variations are best seen in the following graph.

KEY GRAPH

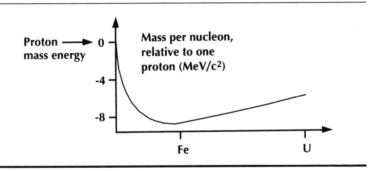

Nuclear binding energy: As the graph shows, the mass of an iron (Fe) nucleus is less than the sum of the masses of the number of neutrons and the number of protons it contains. The masses are given units of MeV/c^2, making use of Einstein's relation $E = mc^2$. The fission reaction given in Key 102 releases about 1 MeV per nucleon times 236 nucleons. The curve can be used to estimate the energy released or absorbed in any nuclear reaction. Combining very light elements like hydrogen and helium to produce heavier ones may release roughly five times as much energy per nucleon as does fission. The true curve has wiggles not shown here, with $^4_2$He lower and $^3_2$He higher than the smooth version. These reveal details of nuclear structure beyond the scope of this Theme.

Key 104 Nuclear fusion

OVERVIEW *Fusion is the sticking together of light nuclei to form heavier ones. It releases more energy per fuel mass than fission but is difficult to achieve under controlled conditions.*

The barrier to fusion: Two nuclei that are only a few nuclear diameters apart repel each other strongly by electrical force but attract only weakly because the strong force diminishes so rapidly with increasing distance. Any reaction based on light nuclei coming together and sticking must include a method for overcoming the electrical repulsion, often called the **Coulomb barrier**. In fusion bombs, explosive compression driven by a fission bomb is used. Crude, but it gets the job done.

Fusion power reactors?

- Controlled fusion of nuclei in numbers great enough to produce heat for electric power generation, but no so great as to melt or explode, has been an elusive goal. In large devices called tokamaks, a hot plasma is held away from the walls of the vessel and compressed by magnetic fields. For fusion of the nuclei in the plasma to occur, the product of their density and temperature must reach a high value. Instabilities that break up the confined plasma have slowed progress.
- In another approach, a small pellet of light atoms is dropped into a vessel into which many laser beams are directed. Just as the pellet reaches the middle, all the lasers fire at it, from every side at once. Heated material expands at the surface, pushing in on the core of the pellet. Densities and temperatures required for fusion are believed to be attainable.

OVERVIEW *Sample questions of the type that might appear on homework assignments and tests are presented with answers.*

Radioactive decay:

- Consider a speck of pure polonium containing 16×10^{23} atoms. It has a half-life of three minutes. How many minutes must we wait to find only 4×10^{23} polonium atoms left? The final number is one quarter of the original number. After one half-life, only one half of the original number of polonium nuclei remain. After another half-life, only one half of that one half still remain. It takes two half-lives or six minutes for the number of polonium nuclei to diminish to one quarter of its original value.

Mass-energy and the nuclear binding curve: It is useful to view the graph in Key 103 while considering the following questions.

- Compare the mass of an atomic nucleus to the sum of the masses of the particles in it. It is different. If it were greater, then the nucleus might fragment into separate particles, with the extra energy appearing as kinetic energy of the fragments. Stable nuclei that do not fragment must have a mass less than the sum of the masses of their constituents.

- Which process, fission or fusion, would release energy from a nucleus of neon? (Neon has atomic mass 20; iron 56.) Fission of neon produces particles with masses totaling more than the neon mass; it could only occur if energy were added. Fusion of two neons would produce a nucleus with less mass than the sum of two neon masses and so could release energy.

- Would fusion of two aluminum (atomic mass 27) nuclei and two neutrons to make iron (atomic mass 56) give more or less energy than fusion of 13 helium (atomic mass 4) nuclei and four neutrons? The 13 helium nuclei and four neutrons have more mass than two aluminum nuclei and two neutrons, so helium fusion would release more energy.

This Theme describes Albert Einstein's 1905 theory of special relativity, which tells us how electric and magnetic fields and particle motion change in situations in which massive particles or observers have speeds close to that of light. It also presents the ideas that Einstein had to work with in order to generalize his theory of relativity to include accelerated frames of reference and, **equivalently**, frames in which gravitational fields exist.

Key 106 Relativity of motion

OVERVIEW *Juggle in your room; juggle in an airplane. You are as good (or bad) at it in one place as in the other. The manipulations needed are exactly the same in both places, even though one whizzes past the other at high speed. The laws of physics are the same in both places and any acceptable physical theory should reflect this fact.*

Frames of reference Key 3 contains a simple prescription for obtaining a description of motion in a new **frame of reference** from that in an original frame. You simply add as vectors the displacement at any time in the original frame and the vector from the origin of the new frame to the origin of the original frame at that time to obtain the displacement at that time in the new frame. The units used to measure length can be changed by multiplying the magnitude of a displacement, velocity, or acceleration by the right scale factor. The origin in time can be shifted also, and the units changed from seconds to years, for example.

Inertial frames: In special relativity only **inertial frames** are considered. If you are in an inertial frame of reference, then all other frames that differ from yours by displacement or a **constant** velocity are also inertial frames. But how do you decide if the original frame is an inertial one? In an inertial frame, a massive object with no forces applied to it remains at rest. If in a trial you see such an object accelerate, your frame is not inertial. Our usual location on the surface of the earth is not inertial, because of earth's gravity and rotation, but for 2D motion in the horizontal plane observed for a time much shorter than a day, we can often neglect the vertical acceleration of gravity and the more subtle effects of rotation without making large errors. Astronauts in free fall in the space shuttle are, for most purposes, in a 3D inertial frame of reference.

Special relativity: The laws of physics are the same in any inertial frame of reference. The simple superposition of motions described in the first section of this Key and Key 3 would not relate observations from different inertial frames if this were not so. We expect the laws of physics to describe motion relative to the origin of an inertial frame and should be suspicious if any law appears to refer to some preferred or absolute frame in which an object at rest is somehow more perfectly at rest than an object at rest in another inertial frame.

Maxwell's equations and the
Michelson-Morley experiment

OVERVIEW *Either the laws of electromagnetism describe
waves that propagate without a medium, or the motion of an
observer through the medium should affect the way light
appears to behave.*

Maxwell's equations and a light wave's speed: James Clerk
Maxwell's equations (Key 77) succinctly describe electric and magnetic phenomena, including electromagnetic radiation. The wave
equation (Key 97) that follows from them predicts the speed of light,
c, in terms of the strengths of the electric and magnetic forces. They
contain no reference to any particular reference frame. No medium
need be present to move as light passes; it consists just of electric and
magnetic fields in a vacuum. But many physicists, uncomfortable
with the idea of waves without a medium, postulated a thin and
imperceptible material, the ether, as the medium in which wave
motion occurs when light passes.

Michelson and Morley's experiment: Michelson and Morley used
fully and partially reflecting mirrors to make a beam of light divide,
follow one of two round trips, and recombine to illuminate a screen.
If the path lengths are nearly equal and monochromatic light (one
wavelength) is used, then an interference pattern appears on the
screen. Different points on the screen correspond to different path-
length differences, so constructive and destructive interference alter-
nate producing a pattern of bright and dark fringes.
- One of the path lengths is set parallel to the direction of the earth's
  velocity, and is set perpendicular. The apparatus can be rotated 90°
  to exchange these two paths. If the ether hypothesis is correct, then
  our velocity through the ether should affect propagation in the parallel and perpendicular directions differently; the fringes should
  shift as the apparatus is rotated. Michelson and Morley saw no significant fringe shift, but the fluctuations in their experiment
  because of vibrations and temperature changes were almost as big
  as the effect they might have expected.

OVERVIEW *If there is no ether, and Maxwell's equations apply as he wrote them in all inertial frames, then the speed of light in vacuum is the same for all observers in inertial frames, whatever their velocities.*

There is no ether: Albert Einstein was more impressed with the beauty and symmetry of Maxwell's equations than with the result of a very noisy experiment. He believed those equations should apply in all inertial frames of reference (it is in this sense that the theory is special, rather than general, and usable in any frame of reference whatever) and inquired what would be the consequences. He believed electromagnetic radiation could propagate in a vacuum by the cooperative action of electromagnetic and magnetoelectric induction (Key 77). These beliefs implied that the speed of light, c, must be the same in every inertial frame of reference, which is the second postulate of Einstein's special theory of relativity.

Constancy of the speed of light: Light emerges from a source that is at rest with respect to you and travels toward you at $c = 3 \times 10^8$ m/s. If the source moves toward you at speed c/2, the light from it travels toward you at speed c. If it moves away at c/2, its light moves toward you at c. If you rocket through your laboratory, in which the source is at rest, at speed c/2. then the light from the source appears to others in the lab to be moving at speed c, and it appears to you to be approaching you in your rocket frame at speed c.

- This is bizarre. Two observers, moving at speed c/2 relative to each other and measuring in the same light wave, both determine the wave speed to be c in their own rest frames. It violates Galilean relativity, the idea that all velocities simply add. But it is a consequence of Einstein's simple and elegant interpretation of Maxwell's equations.

OVERVIEW *In Einstein's special relativity, two events that are simultaneous in one inertial frame of reference are not in another inertial frame that is moving with respect to the first.*

A first consequence of the constancy of c: Let us equip you with a laboratory, a rocket ship, and a light source at rest with respect to the rocket. The light, halfway between the front and back walls of the rocket, flashes briefly as you pass the lab. You see pulses of light strike the front and back walls at exactly the same time. But your lab assistants see the pulses of light moving at speed c **with respect to the lab,** whatever their direction. The rocket moves ahead between the time the light source flashes and the time the pulses of light reach its walls. Your assistants in the lab see light hit the back wall of the rocket before light hits the front wall.

KEY ILLUSTRATION

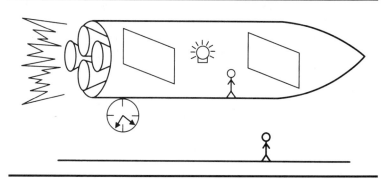

Foolishness? You have never experienced anything like this. Did Einstein fail to find a useful theory? His time differences are small and difficult to measure. Although the loss of absolute simultaneity seems strange and provokes concern about reference-frame invariance of concepts like cause and effect, the special theory of relativity has not at this point predicted a result that we know **from experiments** to be wrong.

OVERVIEW *Time intervals between events and the lengths of rigid objects seem different to observers in different inertial frames, if all see the same speed of light c. Space and time no longer seem to be different kinds of dimensions.*

Time dilation:

- Let us again use your laboratory, assistants, and rocket ship to conduct an experiment involving observers at rest in different inertial frames. This time the apparatus in the rocket is a light source in the center of a screen on one side of the rocket and a mirror on the opposite side. In the rocket you see a flash of light leave the source, travel at speed c across (perpendicular to the direction the rocket is moving relative to the lab) to the mirror, reflect, and travel at speed c back to the screen. If the time it takes the flash to move from the source to mirror is t_0, then the distance from source to mirror is ct_0.

- Observers in the lab see the rocket moving past at speed v, which is smaller than c but may be close to c if we choose to imagine it so. They see the light travel a path of length ct that is the hypotenuse of a right triangle with sides of lengths ct_0 (across the rocket) and vt along the rocket's path. The time for light to cross the rocket, as measured in the lab, is t. Because the path the light travels as viewed from the lab is longer than the path viewed from within the rocket, and the speed of light is c for all observers, the time it takes a flash of light to cross the rocket is longer, measured from the lab, than it is measured from the rocket:

$$t > t_0$$

The quantitative relationship is

$$2t = 2\frac{t_0}{\sqrt{1 - v^2/c^2}}$$

where $2t_0$ is the time between two events at the same place in the frame in which the apparatus is at rest, and $2t$ is the time between the same two events in a frame in which the apparatus appears to move with speed v.

- The apparatus just described could be the heart of a clock. From this we could conclude that a clock that keeps perfect time at rest appears to be running too slowly when it moves by at relativistic speed. This effect is called **time dilation**. Again, this seems very strange, but the time differences are very small for all relative speeds two human observers have achieved. Time dilation has been observed, in the early 1970s, by flying some very accurate clocks in airplanes while others remained on the ground. Nonhuman observers moving at speeds close to c have also demonstrated time dilation: unstable elementary particles that decay with half-life (Key 101) t_0 when observed at rest have half-life $t > t_0$ when accelerated to high speed in a circular particle accelerator.

KEY ILLUSTRATION

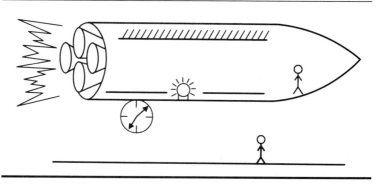

Length contraction: Once again the lab assistants and rocket ship will be most useful. Consider the light clock used to understand time dilation rotated by 90° so the light travels parallel to the direction of the relative velocity between the rocket and the lab. The time dilation will be the same; it cannot depend on details of the apparatus. Two observers, one in the lab and one whizzing by in the rocket, must measure different lengths of the apparatus along the direction of motion. Both see light move back and forth at speed c, and the time it takes seems longer in the lab than in the rocket. A mathematically inclined student can compute the result with a page or so of algebra; he or she must remember that the times for a flash to travel from source to mirror and from mirror to screen, equal in the rocket frame, are not equal in the lab frame. The result is

$$L = L_0 \sqrt{1 - v^2 / c^2}$$

where L_0 is the length of the apparatus in its rest frame and L is the apparent length in a frame in which it moves with speed v. This reduction of the apparent length of a fast-moving object is called the Fitzgerald-Lorentz contraction, or just the Lorentz contraction. Irish physicist G. F. Fitzgerald proposed that all matter moving through the ether contract in the direction of motion, to explain the null result of the Michelson-Morley experiment. Dutch physicist H. A. Lorentz first published the formula just given.

KEY ILLUSTRATION

Space-time: Distances between events, times between events, and lengths of rigid objects all change with the relative speed between the system and an observer. On closer investigation, the distance between two events in one frame is related to both the distance between them and the time between them in another frame. Changing an observer's uniform velocity mixes up spatial and temporal separations in a well-defined, though possibly bewildering way. We live in **space-time,** not space and time separately.

OVERVIEW *High speed contracts lengths and slows clocks. As a consequence, space travel even to other stars many light-years away is possible in principle, but would be a very lonely undertaking for the traveler.*

Biological clocks and time dilation: A person flying by at a relativistic speed would appear to observers on the earth to be living and aging more slowly than a person at rest. This raises the possibility that children and parents could part, later reunite, and find the children older than the parents.

Twins and a high-speed round trip: One decides to take a high-speed cruise through outer space, and the other stays home. The traveler rockets away at nearly the speed of light, suddenly stops and turns around, and rockets back at the same speed. To the twin on the earth, the traveler's clocks—electronic and biological—seem to be running very slowly. When the traveler returns only a few days older than when she left, the twin that stayed behind is many years older.

The paradox: Each twin was in an inertial frame for most of the time of the round trip. Why should one be older at the end? Clocks on the earth must have seemed slow to the traveler. The difference is that the twin at home remained in an inertial frame at **all** times; the traveler accelerated at the beginning, middle, and end of the trip. A detailed description of communication between the twins by flashes of light during the trip clarifies the situation. A good one is in *Conceptual Physics*, eighth Ed., Paul G. Hewitt (Addison-Wesley, Reading, MA, 1989), pp. 45–51.

Space travel: The nearest star, Alpha Centauri, is about four light-years away. To observers on the earth, a round trip made there at almost the speed of light would take eight years. But the rocket's clocks would seem to slow from here. To fast-moving travelers, the distance from here to Alpha Centauri would appear Lorentz-contracted; the trip would take much less than eight years.
  • Given powerful engines and high-energy-density fuel, people could travel to nearby stars and return to earth. But they would return, years older, to an earth centuries older.

Mass increase and mass-energy equivalence

OVERVIEW *In Einstein's special theory of relativity, the apparent mass of an object increases as its speed increases. Its total energy, mc^2, is its rest-mass energy m_0c^2, plus its kinetic energy. Mass is yet another form of energy.*

Mass increase: Einstein's special theory of relativity predicts that an object moving at relativistic speed will appear to be more massive, as well as Lorentz-contracted, and with all its clocks running slow. The formula for the mass increase is similar to those for time dilation and length contraction:

$$m = \frac{m_0}{\sqrt{1 - v^2 / c^2}}$$

m_0 is the rest mass, the mass of the object when it is at rest, and v is its speed when it appears to have mass m. This relativistic mass increase has been observed in the electron and proton beams from high-energy particle accelerators, which give the particles speeds close to c. The momenta of those particles moving at speed c, measured by deflecting the beam with a magnetic field and measuring the deflection angle, are larger than the product m_0v; they are mv with m given by the formula above.

Mass-energy equivalence: The increase of mass with speed is related to the increase in total energy with the addition of kinetic energy. In Einstein's theory, a particle with mass m has total energy mc^2. The same particle at rest has total energy m_0c^2. The difference, $(m - m_0)c^2$, is the kinetic energy of the moving particle. The rest energy of a particle, m_0c^2, suggests that incredibly large amounts of energy could be obtained if we could convert mass to energy; c^2 is a very big number. Partial conversion has been achieved in nuclear fission reactors and in nuclear fission and fusion weapons (Theme 16). Nuclear reactions also generate the heat in the earth's core and power the sun, which slowly loses mass as it shines.

• An early observation that demonstrated the conversion of energy into mass was made by C. D. Anderson, a cosmic-ray physicist in the United States. In 1932, he recorded in a photographic emulsion the tracks left by a pair of particles produced when a cosmic gamma ray was absorbed. Because the experiment was done in a magnetic field, the tracks curved. One was made by an electron, and one was made by a particle with the same charge-to-mass ratio as an electron but a positive charge. This was the discovery of the **positron**—the **antiparticle** to the electron. When any particle and its antiparticle come together, they vanish and their masses vanish, replaced by high-energy photons.

OVERVIEW *Light, which moves in a straight line in an inertial frame of reference, appears to an observer in an accelerating frame to move on a curved path. From this fact, Albert Einstein reasoned that light must be deflected by a gravitation field.*

Light in an accelerating laboratory:

- Imagine yourself in a laboratory in your powerful rocket ship, which is in outer space far from any stars or planets. With all engines off, you are in free fall; if you gently release a baseball in the middle of your lab, it will float in the air before you in a fixed position. If you throw the baseball or turn on a laser, the ball or photons will move in perfectly straight lines across your lab.

- Now turn the engines on and accelerate straight ahead at a rate of 10 ms^{-2}. You will notice that one wall of your lab, which we shall call the floor, pushes ahead, which we shall call up, on you and your baseball with forces equal to your weights on the earth. Throw the ball parallel to the floor, and it appears to fall in a parabolic trajectory towards the floor. Actually, the ball moves in a straight line, ahead (up) with the same component of velocity that the rocket had in that direction at the instant you released it, and parallel to the floor with the component of velocity in that direction that your hand imparted to it (Key 3). But the velocity of the rocket constantly increases, and the floor accelerates up toward the ball. Even though the ball moves in a straight line, inside your lab it appears to fall, as if there were an acceleration of gravity of 10 ms^{-2}, down (Key 4).

- Shine your laser across the lab and the same thing happens. The light moves in a straight line, across and ahead. The acceleration of the floor brings it closer to the light beam during the time the light propagates across the lab. In the lab, the light appears to fall just like the baseball. The distance the light falls is too small to measure because the speed of the light is so great, but if you could command a much greater acceleration of the rocket you could make the fall of the light noticeable.

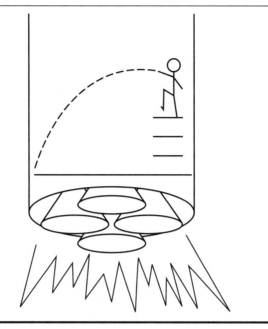

Einstein's equivalence principle: Albert Einstein concluded that there is no experiment you could perform in a small lab to determine whether it was in an accelerating rocket or in the gravitational field of a planet. Just as he assumed the constancy of the speed of light in special relativity and discovered its consequences, he stated this **principle of equivalence** and considered its consequences as he groped toward a general theory of relativity.

Light must fall in a gravitation field: The first consequence of the equivalence principle is simply that light must fall just as matter does in a gravitational field. Its speed is too great for us to detect the fall here on earth, but perhaps near a star or other massive celestial object we could detect gravitational deflection of light. Einstein completed and published his general theory of relativity in 1916 and predicted the angle through which light grazing the surface of the sun should be deflected. In 1919, during a total eclipse of the sun in which the moon passed between the earth and the sun, observers on the earth saw stars beyond the sun and measured their locations precisely. The light from them was deflected by the sun, through just the angle Einstein had predicted.

Key 114 Gravity, time, and space

OVERVIEW *Clocks moving at relativistic speeds appear to run slow. So do clocks that are accelerating rapidly and, by Einstein's equivalence principle, clocks that are in strong gravitational fields. If the speed of light is the same for all observers, and gravity affects the rates of clocks, then gravity must somehow affect space and the lengths of measuring sticks as well.*

Acceleration slows a clock: In one experiment at least, special relativity leads one to expect that an accelerating clock runs slower than the clocks that are not accelerating. Consider two clocks on a merry-go-round, one exactly in the center and one out at the rim. Imagine such a high-powered merry-go-round that the linear speed at the rim is close to the speed of light. Standing on the ground and watching the system spin, you should expect the clock at the rim to run slower than the one at the center, because of the time dilation of special relativity. An observer riding on the merry-go-round would also see the clock at the rim running slower than the one at the center, but would think of both as being at rest. In the rotating frame, there is a very strong centrifugal force at the rim, but none at the center. The rotating observer would conclude that a strong acceleration field causes a clock to slow.

Gravity slows clocks: If acceleration slows a clock that is at rest in one particular accelerated frame of reference, perhaps it does so in any accelerated frame of reference. It is certainly in the spirit of Einstein's equivalence principle to assume that all accelerations have the same effect on the rate of a clock. Equivalence would require, then, that an acceleration of gravity have the same effect. A clock on the surface of the earth must run slower than one in orbit around the earth, slower by a smaller amount than a clock at the top of a tall building. This phenomenon has been observed many times using clocks based on the emission and absorption of photons by atoms and by nuclei.

Red shifts: Consider an atom on the earth's surface that is excited, with its electrons vibrating at a characteristic frequency. It emits a photon with frequency f and energy hf. To an observer many earth radii above the surface, the atoms seems to be vibrating slower than those near the observer do. The frequency of the photon, when it arrives, is lower than that of one emitted by an identical atom near the observer; the wavelength of the photon from the earth's surface is longer than expected classically. If the wavelength corresponds to a green color, it is shifted toward the red from what is expected classically.

- A photon, detected at the top of a tall building, will have a frequency too low and a wavelength too long by an amount that can be computed by combining quantum and relativistic ideas. The photon energy hf will be too small by $(hf/c^2)gh$, where h is the vertical separation between the source and the emitter. Even though gh/c^2 is a very small number for any existing buildings, a very precise nuclear spectroscopy invented in the 1950s by Mossbauer was used by Pound and Rebka in the early 1960s to observe the gravitational red shift produced by a vertical separation of ≈ 10 m. Larger gravitational red shifts have been observed in light from the sun and light emitted close to the surfaces of small, massive celestial objects including white dwarfs and neutron stars.

Gravity affects space: Just as in special relativity, the constancy of the speed of light c in all general frames of reference combines with the fact of gravitational red shifts or time dilations to imply that gravity must affect space as well. It was known before Einstein formulated his general theory of relativity that the elliptical orbit of Mercury behaved in a nonclassical way. Even after the effects of all the other planets were accounted for, there remained an unexplained difference between Mercury's motion and the expected elliptical orbit. The difference is that the ellipse rotates very slowly, rather than remaining fixed in space (perihelion precession, the rotation of the point on the orbit closest to the sun). It was a source of great satisfaction to Einstein that when he used his new theory to calculate Mercury's motion, it predicted a precession of the perihelion point equal to what had been observed.

Mass and the geometry
of space-time

OVERVIEW *Acceleration, whether changing velocity far
from any mass or the gravitational acceleration on an object
at rest near a mass, alters length as well as time. Mass, the
source of gravity, affects space-time.*

Acceleration alters lengths: The merry-go-round of Key 114, equipped
with metersticks rather than clocks, is helpful in generalizing from
special relativity regarding distance as well as time. If the turntable is
at rest, then all the metersticks on it appear to an observer on the
ground to be exactly 1 m long. If the turntable rotates with relativistic
speed at the rim, a meterstick at the rim and parallel to the rim appears
to be Lorentz-contracted. An observer on the merry-go-round and
rotating with it would also see the meterstick near the rim shorter than
one at the center, but would think of the huge centrifugal force at the
rim as the cause. By equivalence, a strong gravitational field should
make a meterstick that is oriented perpendicularly to it appear shorter.

Curved space-time: One way of dealing with altered lengths is to
describe space-time as curved rather than flat, with the degree of curvature larger in regions where the accelerations of gravity are larger.
Flat-space rules such as the Pythagorean theorem and the constant
180° sum of the interior angles of any triangle fail in curved space.
On the surface of a sphere, you can construct a triangle with three 90°
angles and three equal sides, each one quarter of the circumference.
Gravity produces a mathematically similar distortion of space-time.

Mass is the source of the curvature: The source of gravitational fields
is mass. It is the presence of mass that causes the curvature of space-
time. The attraction of two nearby masses can be thought of as a fic-
titious force, like centrifugal force, that is not necessary in a correct
curved frame of reference. The masses move in straight lines in a
space-time curved by their presence.

Key 116 Gravity waves

OVERVIEW *Einstein's general theory of relativity predicts that moving masses emit gravitational radiation, waves that propagate for long distances.*

Gravitational radiation: Just as moving electric charges radiate electromagnetic waves, moving mass can radiate propagating fluctuations in the curvature of space-time—gravity waves. Einstein's general theory of relativity describes these waves in detail. They were not thought of as something we might see or study until the early 1960s, when Joseph Weber at the University of Maryland invented a detector of gravity waves. Since then more and more sensitive detectors have been constructed. But the waves from supernova explosions of stars and other possible sources have proven too weak to detect.

Evidence for gravity waves: There are astrophysical phenomena that constitute indirect evidence for gravity waves. The most compelling is the rate at which the rotation of a pair of compact massive objects has been observed to decrease. One of the objects is a pulsar, which emits bursts of electromagnetic radiation at very regular intervals. Doppler shifts (Key 49) of the times between pulses allow accurate determinations of the rotation rate. In the binary pulsar—discovered by U.S. astrophysicists Joe Taylor and Russell Hulse in 1974 and then observed by them for the next ten years—the rotation is slowing down exactly as would be predicted from the expected rate of energy loss by gravitational radiation.

Black holes, cosmology,
dark matter, and dark energy

OVERVIEW *The curvature of space-time can become
extreme near compact massive objects or when viewed on
the scale of the entire universe. In these realms, classical
physics fails to explain the strange phenomena that occur.*

Black holes: A mass great enough and compact enough can deflect photons (Key 113) strongly enough to hold them in orbit (i.e., prevent
their escape). This happens when orbital speed equals the speed of
light. Neither matter nor light can escape from close to such an
object. J. A. Wheeler invented the term **black hole** in the 1960s to
emphasize this property. There is a growing body of experimental
evidence that has no simpler explanation than the existence of black
holes in our galaxy. Their effect on space-time is to curve it not just
a little, but so much that it splits into two distinct regions, inside and
outside the black hole. In the study of these interesting but distant and
obscure objects, general relativity has become an applied science as
well as an elegant theory.

Cosmology: Does all the matter in the universe curve space-time into a
closed shape, like the interior of a black hole, or is there less total
mass so that the universe is **open**, like the classical picture of a space
that goes on forever? Or is the universe **flat**, just on the edge between
these two states? Much evidence convinces us that the present universe has expanded from a very small, very hot object (the "big
bang") over the last 13.7×10^9 years. It may collapse again, expand
forever, or find a delicate balance between the two, depending on the
answer to the first question posed. General relativity has been crucial
in determining what present-day observations tell us about the earli-
est history of the universe—an exercise that is far from complete.

Dark matter: At the level of galaxies and clusters of galaxies, orbital
speeds far from the object's center have been observed that are too
fast for the orbiting masses to be gravitationally bound by the object's
mass. This is so if the mass is inferred from the visible light emitted
by the stars in the object. Yet galaxies and galactic clusters are stable.
There must be mass in them that is not visible, dark matter that does
not interact with electromagnetic radiation. What kind of particles

carry this mass is not known and is an important question addressed by current research in astrophysics and particle physics.

Dark energy: The universe continues its expansion that began with the big bang. Careful measurements of cosmic microwave background radiation, which ceased interacting strongly with matter early in cosmological time, strongly suggest that the universe is **flat**. If so, at least 90% of the mass and energy (Key 112) in the universe must be dark. Not all the dark mass-energy can be in the form of mass, because observations made since 1990 have established that something is increasing the rate of expansion of the universe, causing galaxies moving away from each other to do so at ever-increasing speeds. What this dark energy may be is not known. Exotic theories like brane theory and string theory, as well as extensions of the **standard model** to include new, dark particles, attempt to produce answers that may be tested experimentally.

OVERVIEW *Sample questions of the type that might appear on homework assignments and tests are presented with answers.*

Constancy of the speed of light:
- Riding in a rocket, you approach a light with a speed of 2×10^8 m/s. How fast does its electromagnetic radiation move past you? Galilean relativity suggests 5×10^8 m/s as a possibility to consider, which is the sum of light speed and the rocket's speed, but special relativity tells us that the speed of any light, viewed by any observer, is 3×10^8 m/s.

Relativistic changes in length, mass, energy, and time:
- A meterstick with a rest mass of 1 kg moves past you rapidly, traveling in a direction parallel to its one-meter-long dimension. Your measurements show it to have a mass of 4.0 kg. What length do your measurements show it to have? The same factor $(1 - v^2/c^2)^{1/2}$ appears both in the formula for mass increase and in the formula for length contraction. If mass increases by a factor of four then length contracts to one fourth of the rest length, or 25 cm.
- To what speed must you accelerate a particle to make its total energy be three times its rest-mass energy? The total mass-energy is the rest mass-energy multiplied by the factor given in the previous paragraph. So what is v if

$$\frac{1}{\sqrt{1 - v^2 / c^2}} = 3$$

It must be that

$$1 - \frac{v^2}{c^2} = \frac{1}{9} \text{ and}$$

$$\frac{v^2}{c^2} = \frac{8}{9} \text{ and so}$$

$$v = c \times 0.94$$

Gravitational red shift: Green light is emitted from the surface of a massive star. It comes to you in your spaceship orbiting far from the star. Has its color changed? From your point of view, a clock on the surface must appear to run slow because of the strong gravitational field there. The vibrating atom that emitted light must seem to be vibrating more slowly than it would if it were in your spacecraft. Thus the light that comes to you from the star's surface will be lower in frequency than if it were emitted from a source near you; it is red-shifted.

A black hole:

- Around the earth there are communications satellites in stable orbits. Around a small enough, heavy enough object, light could be put into a stable orbit: true or false? True: The equivalence principle alone suggests the possibility, and detailed calculations with Einstein's theory predict its occurrence.

GLOSSARY

*Included here are the definitions of many, but not all, of the terms used in the Keys. Terms printed in **boldface** in the definitions are further explained under their own entries in this glossary. For terms not listed here, please consult the index.*

Absorption   The transfer of **energy** from a light **wave** to matter.

AC   Alternating **current**; electrical signals that vary **harmonically** in time.

Acceleration   A **vector** describing the rate of change of **velocity**.

Air friction   The **force**, which increases with increasing speed, exerted by air on any object moving through.

Alpha ray   A He **nucleus** with very high kinetic **energy**, often produced by a nuclear reaction.

Ammeter   An instrument used to measure electric **current**.

Amplitude   The size of **field**, of the swing of a **pendulum**, or of the disturbance created by a passing **wave**.

Angular acceleration   A **vector** describing the rate of change of an **angular velocity**; it may describe either a change with time of the orientation of the rotation axis or of the **angular speed**.

Angular momentum   A **vector** analogous to linear **momentum**, the product of **angular velocity** and the **moment of inertia**.

Angular speed   A **scalar** describing the rate of rotation of an object.

Angular velocity   A **vector** describing the rate of rotation of an object and the orientation in space of the rotation axis.

Antinode   A place in a standing wave where the **amplitude** is a maximum.

Antiparticle   An elementary particle with the same **mass** but opposite electrical **charge** to that of its related particle.

Atom   The smallest unit into which an **element** can be divided without losing its identity as, for example, aluminum, oxygen, or iron.

Atomic mass   In this **mass unit**, the **mass** of an **atom** is approximately equal to the number of **neutrons** and **protons** in its **nucleus**.

Atomic number   The number of **protons** in the **nucleus** of an **atom**, and also the electrical **charge** of the **nucleus** in **units** of the magnitude of the **electron's charge**.

Beta ray   An **electron** with a large kinetic **energy**, usually emitted from a **nuclear** reaction.

Black body   An object that absorbs all electromagnetic **radiation** falling on it and **emits** as much **energy** as it absorbs.

Black-body spectrum   The distribution over either **wavelength** or frequency of the **power** emitted by a **radiating black body**.

Bohr atom   Niels Bohr's model formulated at the beginning of quantum mechanics that first predicted discrete, rather than continuous, atomic **spectra**.

Buoyant force   The upward force exerted in a gravitational **field** by a liquid on an immersed object.

Carnot efficiency   The greatest efficiency a **heat engine** could possibly have; a theoretical limit.

Center of mass   The one point in or near a rigid object that moves as if all its **mass** were concentrated at the point.

Charge   A measure of the quantity of electricity in an object.

Coherent   Happening all together, as in the movement of the legs of the marching soldiers or in a **laser** in which separately **emitted** light **waves** line up, crest to crest.

Complementarity   Niels Bohr's principle that **duality** is central to quantum mechanics, rather than a problem that would someday be solved.

Compound   A material containing more than one **element** in specific proportions.

Concave   Curved the way a cave or bowl is, with the center further away from you than the edges.

Conduction   The flow of **heat** or electric **current** through matter.

Conductor   Matter capable of supporting the flow of **heat** or electric **current**.

Constructive Contributions adding with similar signs, producing a sum larger than any one contribution, as in **constructive interference** of **waves**.

Convection **Heat** flow in a vapor or liquid by the movement of hot masses.

Convex Curved the way a hill or ball is, with the center closer to you than the edges.

Cosmology The science that studies the origin, present state, and possible futures of the entire universe.

Crystal A solid in which **atoms**, **molecules**, or clusters of them are arranged in a perfectly repeating pattern.

Current The quantitative measure of the rate of flow of some quantity such as electrical **charge**, **heat**, or **fluid mass**; the amount of electrical **charge**, **heat**, or **fluid mass** passing a point in 1 s.

DC Direct **current**; electrical signals that keep a constant sign in time.

de Broglie wavelength The **wavelength** of the quantum mechanical wave associated with any object that has **momentum**.

Destructive Contributions adding with different signs, producing a sum not much larger than any one contribution, as in **destructive interference** of **waves**.

Diffraction The **wave** phenomenon in which spreading and **interference** occur after the **wave** passes an obstacle.

Diffraction grating A fine grid of alternating transparent and opaque lines that strongly **diffracts** light.

Dispersion Spreading out, usually in a spatial dimension, of **waves** in order of their **wavelengths**.

Doppler shift The frequency change that occurs between a **wave** source and an observer with non-zero relative velocity.

Duality The principle that quantum objects behave as **waves** in some experiments, as particles in others.

Electromagnet An electric circuit designed to produce a magnetic **field** when **current** flows in it.

Electromagnetic induction   The electrical effect produced by a magnetic **field** that is changing in time.

Electromotive force (EMF)   A non-Coulomb **force** on a **charged** particle; an inductive **force**, for example.

Electron   An elementary particle with negative **charge** and small **mass**; responsible for the size and chemical properties of atoms.

Electroscope   An instrument for measuring electric **charge**.

Element   Any material that does not split into simpler constituents by chemical reaction.

Emission   The process in which an excited **atom radiates** light.

Energy   A conserved quantity discovered in mechanics after Newton's work, and later generalized to non-mechanical forms as well.

Entropy   A measure of the degree of disorder in a **thermodynamic** system.

Equilibrium   A state of tending to remain at rest or to persist without change.

Ferromagnetic   Strongly magnetic, like iron, nickel, and cobalt.

Field   A **scalar** or **vector** quantity defined at every point in a system; **temperature** or the **acceleration** produced by gravity, for example.

Fluid   Matter that can flow; a liquid or a gas.

Flux   The amount of stuff crossing a unit area in a unit time, useful in describing the flow of conserved material and electric, magnetic, or gravitational **fields**.

Force   A **vector** describing the direction and magnitude of a push.

Frame of reference   A real or hypothetical seat from which an observer watches objects move.

Free fall   Motion of an object acted on by no **forces** other than gravitational ones.

Friction   A **force** present only when an object moves while in contact with another object, and which transforms **mechanical energy** into **heat**.

Galvanometer An instrument for measuring electrical **current** by the deflection of a **current**-carrying wire in a magnetic **field**.

Gamma ray Electromagnetic **radiation** with frequencies higher than those of **X rays** and light.

Generator A machine that produces electric **potential difference** and **current** by the **forces** on **electrons** in wires that move through magnetic **fields**.

Glass A dense solid lacking the long-range order of a **crystal** or a **quasicrystal**.

Half-life The time it takes for half of any starting collection of identical, unstable **nuclei** to decay.

Heat **Energy** that flows from one object to another because their **temperatures** differ.

Heat engine A machine that converts some of the **heat** flowing into it to **mechanical work**; can convert **mechanical work** to **heat** if designed to run the other way.

Heat of fusion The **heat** required to convert a unit **mass** of solid to liquid.

Heat of vaporization The **heat** required to convert a unit **mass** of liquid to vapor.

Impulse The product of a **force** and the time for which it acts, this **vector** is equal to the change in **momentum** produced by the **force**.

Incoherent Not **coherent**, random.

Inertia The property of resisting changes in **velocity**.

Inertial frame An observer's point of view from which objects not acted on by **forces** are seen to move with constant **velocities**.

Insulator An object that does not **conduct**.

Intensity **Energy flux** A useful measure of the strength of a propagating **wave**.

Interface A surface that is the boundary between two distinguishable regions.

Interference   The adding and canceling of **amplitudes** that occurs when two or more **waves** come together.

Ionizing radiation   Electromagnetic **radiation** or energetic **charged** particles that knock **electrons** out of **atoms** when passing through matter.

Isotope   **Nuclei** with the same number of **protons** but different numbers of **neutrons** are different **isotopes** of an **element**.

Kepler orbit   The elliptical **trajectory** of a planet around a star at one of the foci of the ellipse.

Laser   A source of **coherent** light.

Lens   A piece of **glass** with smooth but curved surfaces that deflects light passing through it through angles that depend on the locations of the light **rays**.

Light-year   The distance light travels in a time of one year.

Longitudinal   The **wave polarization** in which the **wave** motion is **parallel** to the propagation direction.

Mass   The measure of **inertia**.

Mechanical equivalent of heat   The amount of **mechanical work** that, when done against **frictional forces**, produces one standard **unit** of **heat**.

Mechanics   The science that describes the motions of **massive** objects and how **forces** produce their motions.

Medium   That which moves when a **wave** passes through it.

Molecule   The smallest piece of matter that has the same properties as larger amounts; one or many **atoms** bound together.

Moment of inertia   The measure of an object's **rotational inertia**.

Momentum   A **vector** that is the product of the **mass** and velocity of an object.

Motor   An electromagnetic machine for converting electrical **energy** to **mechanical work**, usually by producing **torque** that causes a shaft to rotate.

Natural frequency   The frequency of an **oscillator** moving free from any external interference.

Neutron   A **massive**, un**charged** particle found in **atomic nuclei**.

Node   A place in a standing **wave** where the **amplitude** is zero.

Nuclear binding energy   The **energy** required to separate the **neutrons** and **protons** in a **nucleus** from each other.

Nuclear force   The **force** that holds the **neutrons** and **protons** in a **nucleus** together.

Nucleon   A **neutron** or **proton**.

Nucleus   The small, positively **charged** center of an **atom** that contains most of its **mass**.

Orbit   The closed path of a planet or **satellite**.

Oscillation   A motion that repeats periodically.

Parallel   Having the same direction.

Pendulum   A **mass** supported on a rod or string near the earth's surface that is free to swing.

Period   The time in which exactly one cycle of an **oscillation** occurs.

Periodic oscillating   Repeating at regular intervals.

Perpendicular   Having directions 90° apart.

Photon   One quantum of light.

Planck's constant   The small number that sets the scale (length, time, energy) below which quantum effects are important.

Polarization   The direction of **wave** motion, relative to the propagation direction.

Pole   A concept in magnetism analogous to **charge** in electricity.

Positron   The **antiparticle** to the **electron**; has the same **mass** but positive **charge**.

Potential   The electrical quantity that describes the electrical potential **energy** per unit **charge** that a **charge** would have if placed at the point described.

Potential difference   The difference between the electrical **potentials** at two different points.

Power   The rate at which **energy** is transferred.

Pressure   The **force** per unit area exerted by a fluid on its container or on any immersed object.

Prism   A piece of **glass** with smooth, flat surfaces that are not **parallel**.

Proton   A **massive charged** particle found in **atomic nuclei**; the hydrogen **nucleus**.

Quasicrystal   A solid structure intermediate between a **crystal** and a **glass**.

Radiation   Propagating **waves**, usually electromagnetic.

Radioactivity   The emission of high-**energy photons** or **charged** particles by some unstable **atomic nuclei**.

Rarefaction   Expansion, opposite to compression.

Ray   A thin beam of light with a well-defined direction of travel.

Reaction force   The Third-Law **force** produced by a pushed object that acts back on the pusher.

Reciprocity   The idea that a machine or a process can operate in the reverse direction to that first considered.

Refraction   The change of direction of light **rays** crossing an **interface** from one transparent **medium** into another.

Resistor   An electronic circuit component whose operation is described by Ohm's law.

Resonance   An **oscillator's** large-**amplitude** response to a **force** applied **periodically** at or near the **oscillator's natural frequency**.

Rotational inertia   The property of resisting any change in the rate of rotation of an object.

Satellite   Any natural or artificial object **orbiting** around a planet or star.

Scalar   A physical quantity that has magnitude and **units**, but not direction; temperature, for example.

Scientific notation   The use of powers of ten to express compactly very large and very small numbers.

Semiconductor   A material through which **electrons** can move freely as in a **conductor**, but with relatively few such **electrons** (none at low **temperatures** in pure **semiconducting** materials); widely used in electronics.

SI   Système International, the name for the standard **units** used by physical scientists.

Simple harmonic motion   Motion like the projection onto a vertical screen of the shadow of an object moving in a horizontal circle before the screen; the position is a **sine** or cosine function of time.

Sine wave   A mathematical function that describes (1) the position of a **pendulum** as a function of time, and (2) some other **periodic** motions, including some simple **waves**.

Solenoid   A helical **electromagnet**.

Space-time   The four-dimensional frame of reference in which special and general relativity are considered.

Spectrograph   An optical instrument in which a grating or prism sends light of different **wavelengths** in different directions and photographic film records the **intensity** as a function of position (**wavelength**).

Spectrometer   An optical instrument in which a rotatable grating or prism sends light of one selected **wavelength** to a detector.

Spectrum (pl. spectra)   A graph of **intensity** (vertical axis) as a function of **wavelength** (horizontal axis) or frequency or **energy**.

Stability   The tendency of a system to remain in its present state rather than changing.

Standing wave   A **superposition** of two or more propagating **waves** that produces a pattern that **oscillates** in **amplitude** but does not otherwise move; the motion of a string on a stringed musical instrument, for example.

Superconductor   An electrical **conductor** that, at low enough **temperature**, has exactly zero electrical resistance.

Superposition   Making a new **wave** or a new motion by simultaneously producing two or more simple **waves** or motions.

Supersonic   Having a speed in a **medium** greater than the speed of sound in it.

Temperature   The measure of the tendency for **heat** to flow from one object to another.

Tension   In a solid, the **forces** tending to pull its molecules closer together.

Thermal expansion   The increase in size (length, volume) that occurs in most condensed matter with increasing **temperature**.

Thermodynamics   The science that studies **heat** flow, **temperature** changes, and the associated changes of other physical properties.

Torque   The product of a **force** and the **perpendicular** distance from its line of action to a rotation axis; tends to change the rate of rotation of the object to which it is applied.

Trajectory   The path followed by a moving object.

Transformer   Two or more coils of wire arranged so the magnetic **field** produced by a **current** in one links the turns of the others, often wound around a soft iron core; by **electromagnetic induction AC** electrical signals pass between the coils.

Transverse   **Perpendicular** or sideways; **transverse polarization** of a **wave** is **perpendicular** to its propagation direction.

Uncertainty   In quantum mechanics, the impossibility of simultaneously determining the exact values of all physical variables; the position and **momentum** of a **massive** particle, for example.

Unit   The standard of comparison used to measure a quantity; meter, kilogram, or second, for example.

Vector   A quantity that has direction as well as magnitude.

Voltage   Electric **potential**.

Voltmeter   An instrument used to measure electrical **potential difference** between two points in an electric circuit.

Wave   A disturbance in a **medium** that propagates through the **medium**.

Wave equation   A mathematical description of a **wave** that relates its temporal and spatial rates of change.

Weight   The gravitational **force** acting on an object near a star or planet.

Work   The product of **force** and the distance moved **parallel** to the force; equal to the **energy** transferred to the object moved.

X ray   Short-**wavelength**, high-**energy** electromagnetic **radiation**; **wavelengths** are typically 10^{-10} m.

INDEX

Success on Advanced Placement Tests Starts with Help from Barron's

Each May, thousands of college-bound students take one or more Advanced Placement Exams to earn college credits—and for many years, they've looked to Barron's, the leader in Advanced Placement test preparation. You can get Barron's user-friendly manuals for 20 different AP subjects, most with optional CD-ROMs. Every Barron's AP manual gives you—

- **Diagnostic tests**
- **Extensive subject review**
- **Full-length model AP exams**
- **Study help and test-taking advice**

Model exams are designed to reflect the actual AP exams in question types, subject matter, length, and degree of difficulty. All questions come with answers and explanations. All books are paperback.

AP: Art History, w/optional CD-ROM
Book: ISBN 978-0-7641-3737-2, 448 pp.
Book w/CD-ROM: ISBN 978-0-7641-9463-4

AP: Biology, 2nd Ed., w/optional CD-ROM
Book: ISBN 978-0-7641-3677-1, 496 pp.
Book w/CD-ROM: ISBN 978-0-7641-9327-9

AP: Biology Flash Cards
ISBN 978-0-7641-7868-9, 504 cards

AP: Calculus, 9th Ed., w/optional CD-ROM
Book: ISBN 978-0-7641-3679-5, 672 pp.
Book w/CD-ROM: ISBN 978-0-7641-9328-6

AP: Calculus Flash Cards
ISBN 978-0-7641-9421-4, 300 cards

AP: Chemistry, 4th Ed., w/optional CD-ROM
Book: ISBN 978-0-7641-3685-6, 688 pp.
Book w/CD-ROM: ISBN 978-0-7641-9329-3

AP: Chemistry Flash Cards
ISBN 978-0-7641-6116-2, 500 cards

AP: Computer Science Levels A and AB, 4th Ed., w/optional CD-ROM
Book: ISBN 978-0-7641-3709-9, 736 pp.
Book w/CD-ROM: ISBN 978-0-7641-9350-7

AP: English Language and Composition, 2nd Ed., w/optional CD-ROM
Book: ISBN 978-0-7641-3690-0, 224 pp.
Book w/CD-ROM: ISBN 978-0-7641-9330-9

AP: English Literature and Composition, 2nd Ed., w/optional CD-ROM
Book: ISBN 978-0-7641-3682-5, 416 pp.
Book w/CD-ROM: ISBN 978-0-7641-9331-6

AP: Environmental Science, 2nd Ed.
ISBN 978-0-7641-3643-6, 688 pp.

AP: European History, 4th Ed., w/optional CD-ROM
Book: ISBN 978-0-7641-3680-1, 336 pp.
Book w/CD-ROM: ISBN 978-0-7641-9332-3

AP: French w/Audio CDs, 3rd Ed., w/optional CD-ROM
Book w/3 audio CDs: ISBN 978-0-7641-9337-8, 464 pp.
Book w/3 audio CDs and CD-ROM: ISBN 978-0-7641-9336-1

AP: Human Geography, 2nd Ed.
ISBN 978-0-7641-3817-1, 336 pp.

AP: Italian Language and Culture, w/Audio CDs, 2nd Ed.
ISBN 978-0-7641-9368-2, 448 pp.

AP: Micro/Macro Economics, 2nd Ed.
ISBN 978-0-7641-3361-9, 312 pp.

AP: Physics B, 4th Ed., w/optional CD-ROM
Book: ISBN 978-0-7641-3706-8, 476 pp.
Book w/CD-ROM: ISBN 978-0-7641-9351-4

AP: Physics C, 2nd Ed.
ISBN 978-0-7641-3710-5, 768 pp.

AP: Psychology, 3rd Ed., w/optional CD-ROM
Book: ISBN 978-0-7641-3665-8, 320 pp.
Book w/CD-ROM: ISBN 978-0-7641-9324-8, 320 pp.

AP: Spanish w/Audio CDs, 6th Ed., and w/optional CD-ROM
Book w/3 audio CDs: ISBN 978-0-7641-9406-1, 448 pp.
Book w/3 audio CDs and CD-ROM: ISBN 978-0-7641-9405-4

AP: Statistics, 4th Ed., w/optional CD-ROM
Book: ISBN 978-0-7641-3683-2, 560 pp.
Book w/CD-ROM: ISBN 978-0-7641-9333-0

AP: Statistics Flash Cards
ISBN 978-0-7641-9410-8, 500 cards

AP: U.S. Government and Politics, 8th Ed., w/optional CD-ROM
Book: ISBN 978-0-7641-3820-1, 510 pp.
Book w/CD-ROM: ISBN 978-0-7641-9404-7

AP: U.S. History Flash Cards
ISBN 978-0-7641-7837-5
Boxed set of 500 3" x 5" cards

AP: United States History, 8th Ed., w/optional CD-ROM
Book: ISBN 978-0-7641-3684-9, 496 pp.
Book w/CD-ROM: ISBN 978-0-7641-9334-7

AP: World History, 8th Ed., w/optional CD-ROM
Book: ISBN 978-0-7641-3822-5, 512 pp.
Book w/CD-ROM: ISBN 978-0-7641-9403-0

AP: World History Flash Cards
ISBN 978-0-7641-7906-8, 400 cards

Barron's Educational Series, Inc.
250 Wireless Blvd.
Hauppauge, NY 11788
Call toll-free: 1-800-645-3476
Order by fax: 1-631-434-3217

In Canada:
Georgetown Book Warehouse
34 Armstrong Ave.
Georgetown, Ontario L7G 4R9
Canadian orders: 1-800-247-7160
Order by fax: 1-800-887-1594

Please visit **www.barronseduc.com** to
view current prices and to order books

(#93) R1/08